POCKET NATURE

SEASHORE

POCKET NATURE

SEASHORE

CHRIS GIBSON

DORLING KINDERSLEY

LONDON, NEW YORK, MUNICH,
MELBOURNE, AND DELHI

DK LONDON
Senior Art Editor Ina Stradins
Senior Editor Angeles Gavira
Editor Miezan van Zyl
DTP Designer John Goldsmid
Picture Editor Liz Moore
Illustrators Jane Durston, Andy Mackay
Production Editor
Jonathan Ward
Managing Art Editor Phil Ormerod
Managing Editor Sarah Larter
Publishing Manager Liz Wheeler
Art Director Bryn Walls
Publishing Director Jonathan Metcalf

DK DELHI
Designers Arunesh Talapatra
Neha Ahuja, Neerja Rawat
DTP Designers Balwant Singh,
Tarun Sharma
Editors Dipali Singh, Aditi Ray,
Aakriti Singhal, Alicia Ingty
Art Director Shefali Upadhyay
Head of Publishing Aparna Sharma

First published in Great Britain in 2008 by
Dorling Kindersley Limited
80 Strand, London WC2R 0RL

A Penguin Company

A CIP catalogue record for this book is available from the British Library

ISBN 978 1 4053 2862 3

Reproduced by Colourscan, Singapore
Printed and bound by Sheck Wah Tong
Printing Press Ltd, China

see our complete catalogue at
www.dk.com

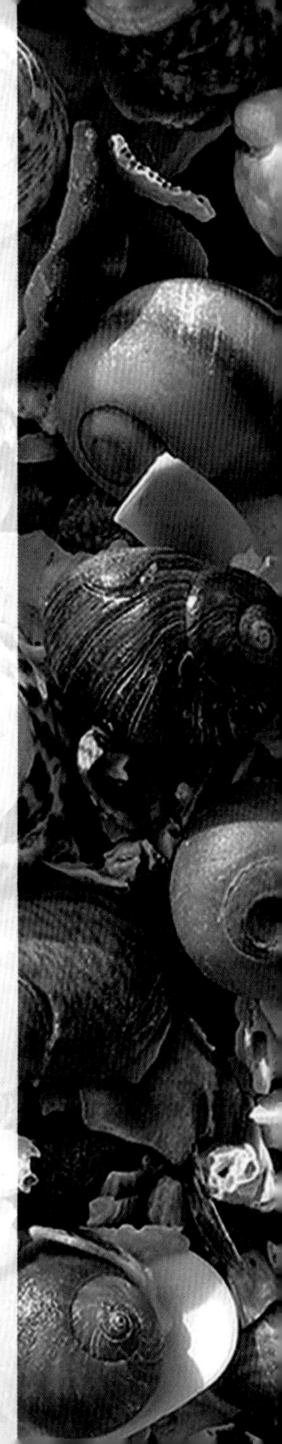

CONTENTS

How this book works

This guide covers over 300 of some of the most commonly found plant and animal species that inhabit the coastal fringes of Europe between the Arctic and the Mediterranean. Some species are widespread, others are localized, but all are typical of coastal habitats, whether rocky shores, cliffs, dunes, shingle, salt marshes, or lagoons. The book is divided into six chapters: higher plants, which covers wild flowers, grasses, trees, and shrubs; lower plants, which includes algae, mosses, and ferns; invertebrates; vertebrates, excluding the birds, which have their own chapter, and strandline finds. The scope of the book generally stops at the low-water mark; most animals or plants below this level can be seen only by divers, or from boats, but a small selection that may drift ashore is presented in the Strandlines chapter.

SCIENTIFIC NAME

CHAPTER HEADING

COMMON NAME

HABITAT
This describes the natural habitats where the species may be found.

BREEDS *on low-lying northern coasts and islands with rocky shores and weedy bays. Winters at sea, often in sandy bays and over mussel beds.*

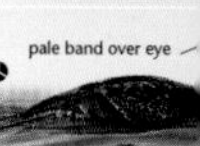

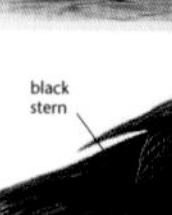

Higher plants

The evolution of the seed (with its protective covering) allowed plants to colonize dry land, and there are relatively few aquatic species of higher (or seed-bearing) plant, especially in the salt-water environment. However, many plants are adapted to withstand salty, coastal environments. Some of the plants found on the coast also grow inland, but the species on the following pages are particularly associated with coastal habitats. The plants are primarily arranged by flower colour – yellow, white, pink and red, blue, purple, and green; the trees and shrubs feature at the end.

SEA WORMWOOD | THREE-HORNED STOCK | SEA CLUB-RUSH | AFRICAN TAMARISK

◁ CHAPTERS
Each chapter opens with an introductory page that gives some general information on the species featured in the pages that follow.

ANNOTATION
Characteristic features of the species are picked out in the annotation.

DETAIL PICTURES
The tinted boxes show variations, including some subspecies, tracks and signs, or simply close-ups of individual parts, such as leaves, flowers, and fruit.

SCALE DRAWING
This gives an indication of the size of the species. See panel right.

DATA BOX
Key information relating to the species is encapsulated in a data box. The categories of information are individually tailored for each group of plants or animals, so that the most appropriate details are given for each. Two constant categories are Distribution and Similar species. Similar Species lists species that look similar to the one featured, often providing a distinguishing feature to help tell them apart. Distribution gives information only about the coastal distribution of the species.

VOICE *Male gives sensuous deep growls and a mechanic*
FEEDING *Molluscs, especia invertebrates gathered during*
DISTRIBUTION *Breeds: Atlar Adriatic; Baltic. More widespr*
SIMILAR SPECIES *Common S and Steller's Eider* (Polysticta *Norway and locally in the Bal*

▽ SPECIES ENTRIES

Each species entry has a description of the species and a main photograph taken in the species' natural setting in the wild. Annotations, scale artworks, and a data box complete the entry.

CLASSIFICATION

This classifies the species at a level that best helps to identify and characterize it: Higher plants: Family; Lower Plants: Phylum; Fungi and Lichens: Kingdom; Algae: Class; Non-arthropod invertebrates: Phylum. Crustacea: Subphylum; Spiders: Class; Insects: Order. Birds: Family; Fish: Family; Reptiles and Amphibians: Class.

llissima (Anatidae)

tirely marine duck with a characteristic
ıead, the Eider is usually easy to identify.
s boldly pied black and white, with green
ıead and a pink flush on its breast. Females
ımage with close dark bars that provide
age on the nest. Highly sociable, Eiders
e rafts offshore, but they are equally familiar
rocks.

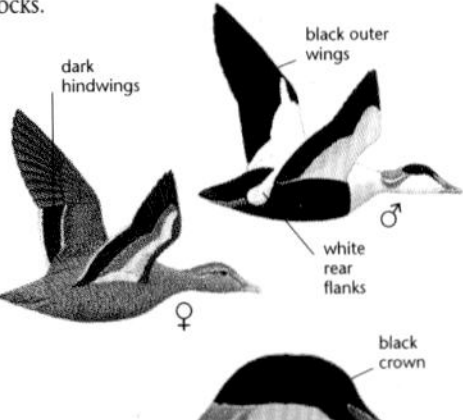

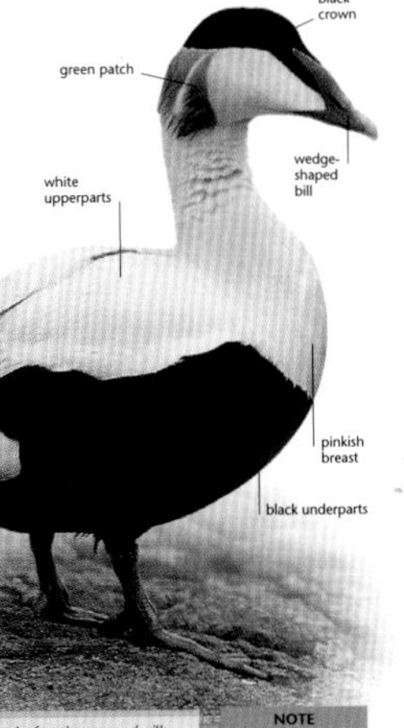

ooh; females respond with
k.
ustaceans, and other
e.
ı France, recently in northern
f range in winter.
King Eider (S. spectabilis)
e Arctic, which winter off

NOTE

Eiders use neck and head muscles to pull mussels from rocks. These are swallowed whole, crushed in the gizzard, and the shell remnants are produced as a pellet.

SCALE DRAWINGS

Two small scale drawings are placed next to each other in every entry as a rough indication of species size. The drawing of the book represents this guide, which is 19cm high. The colour illustration or silhouette represents the species featured in the entry.

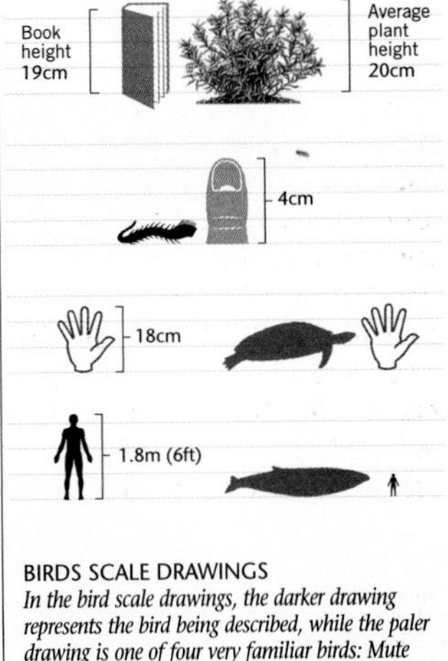

BIRDS SCALE DRAWINGS

In the bird scale drawings, the darker drawing represents the bird being described, while the paler drawing is one of four very familiar birds: Mute Swan, Mallard, Pigeon, and House Sparrow. Sizes below are the length from tip of tail to tip of bill.

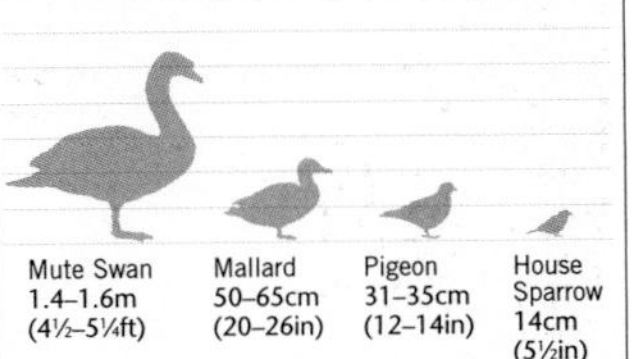

SYMBOLS

Symbols indicate sex, age, season, or view. If an entry has no symbols, it means that the species exhibits no significant differences in these.

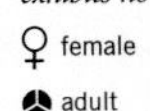

♀ female ♂ male

adult immature juvenile

spring summer autumn

winter

upperside view underside view

DISTRIBUTION

Covers the European coastline, along the shores of the Atlantic (including the Irish and North Seas), Baltic, and Mediterranean.

Life by the sea

The coastline of Europe is very diverse in form and nature – a result of many factors, including the underlying geology and the range of climatic zones it spans, from the arctic to the warm temperate. The unifying factor in this geological diversity is the presence of the sea, which influences the coastline and the wildlife to be found there in several ways.

Energy

Waves, tides, and currents make the sea a very dynamic entity. This dynamism erodes the land, transports eroded material to other areas, and deposits it in suitably sheltered places.

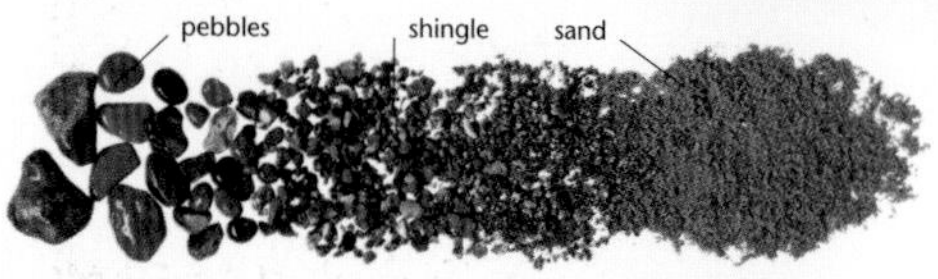

DEPOSITION *is not uniform; currents sort transported material into different sizes: boulders, shingle or pebbles, sand, and silt.*

Salinity

The salinity of sea water makes it impossible for many animals and plants to survive. Those which thrive in salty conditions often show distinct adaptations to cope with it. Brackish water, or water with fluctuating salinity as in estuaries, require wildlife to have even more specialized adaptations.

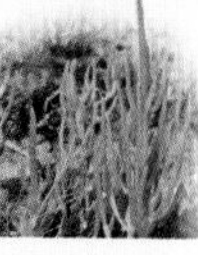

GLASSWORT *has succulent, cylindrical stems that store fresh water for later use.*

SEA-PURSLANE *is covered in crystalline salt deposits that are excreted by salt glands on its leaves.*

Thermal Transfer

A warm water current, the North Atlantic Drift brings in heat from the tropics, making the European coastline significantly warmer than coastlines at equivalent latitudes elsewhere. This allows some southern species to occur further north.

SEA DAFFODIL *is a warmth-loving plant, but extends up the Atlantic seaboard to Brittany.*

MANED SEA-HORSE *is a subtropical fish, found as far north as the English Channel.*

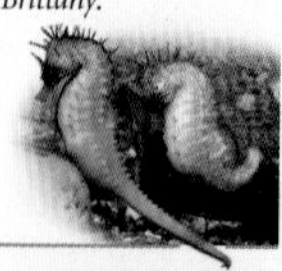

Thermal Buffering

The sea acts as a storage heater – coastal areas are warmer in winter and cooler in summer than the nearby land. Coasts are thus less prone to frost and snow. Temperature differences also lead to the generation of sea breezes, and a higher frequency of fog and mist.

Seas of Europe

The three main sea areas of Europe – the Mediterranean, Atlantic (including the Irish Sea and North Sea), and Baltic – are very different from one another. Each has different environmental features, which greatly influence the wildlife found in and around them. Not surprisingly, several of the plants and animals featured in this book are restricted to just one of the sea areas.

Atlantic

Part of a larger oceanic area, the northeastern Atlantic is highly energetic, with powerful rollers that bring warmth from the tropics, and cause erosion. It has a large tidal range, with well-developed intertidal habitats. Often turbid, the suspended sediment limits the passage of light, which is essential for plant life.

SOUTHERN ENGLISH COAST

GANNETS *breed on islands and cliffs in the north Atlantic, but wanders more widely in search of fish.*

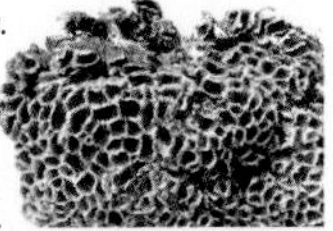

HONEYCOMB WORM *forms extensive reefs that withstand wave attack.*

Mediterranean

An almost enclosed, relatively small body of water, the Mediterranean is microtidal, and so has a reduced intertidal zone. Its warm climate leads to water evaporation, concentrating salt at levels greater than that found in the oceans. Its high salinity and low levels of sediment lend it its characteristic clear azure-blue colour.

BALEARIC ISLANDS

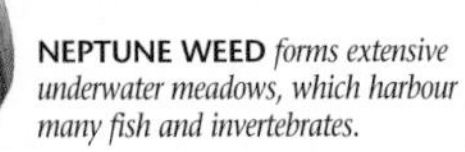

NEPTUNE WEED *forms extensive underwater meadows, which harbour many fish and invertebrates.*

BLACK SEA-URCHINS *graze algae on rocks in the warm Mediterranean waters.*

Baltic

Semi-enclosed and microtidal, the Baltic receives inflows of fresh water from rivers. Its low salinity, shallow nature, and northerly location make it prone to freezing in the winter, and result in a lower and very different biodiversity than that found in the Mediterranean or Atlantic.

CASPIAN TERNS *breed in summer on low, rocky islets in the southern Baltic.*

ALAND ISLANDS

BALTIC TELLIN *is very tolerant of low salinity, and is one of the most abundant invertebrates found on the Baltic coast.*

Habitats

All the plants and animals covered in this book are characteristic of coastal habitats, with many found in a restricted habitat. Each of the main habitats is briefly described below, with a few examples of the wildlife particularly associated with them. Habitat is always a good place to start when attempting to identify unfamiliar species. But preferences are not absolute – mobile species, in particular, can often cause confusion by turning up at an unexpected place.

Rocky Cliff

Near-vertical rocky faces provide nest sites for birds, provided the ledges are wide enough to hold the eggs or provide a base for a nest, and are inaccessible to many predators. The same ledges provide niches for plants, which grow luxuriantly due to the fertilizing effect of bird droppings.

GUILLEMOT *eggs are sharply pointed, meaning they roll in a cricle rather than off the edge of the narrow ledges on which the birds nest.*

ROCK SAMPHIRE *is succulent and cushion-shaped to tolerate salt spray and wind.*

Soft Cliff

Many plant colonizers of soft cliffs (a usually unstable habitat) are able to rapidly colonize new landslips with spreading rhizomes or stolons. Spring-lines provide a habitat for marshland species, while bare cliff faces are favoured by many ground-nesting insects.

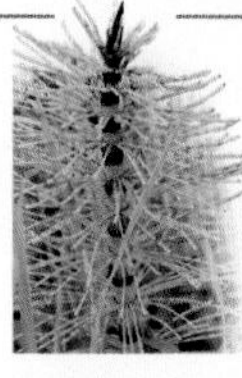

GREAT HORSETAIL *inhabits seepage zones, where springs issue from the cliffs.*

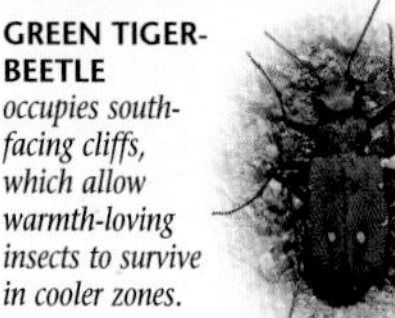

GREEN TIGER-BEETLE *occupies south-facing cliffs, which allow warmth-loving insects to survive in cooler zones.*

Rocky Shore

Wildlife on an intertidal rocky shore must be able to tolerate exposure to air as well as inundation and wave attack. Animals may move with the tide to favoured locations, while others have hard shells for protection. This leads to a distinct zonation of plants and animals.

TURNSTONES *feed on rocky shores, searching for invertebrates under seaweeds and stones.*

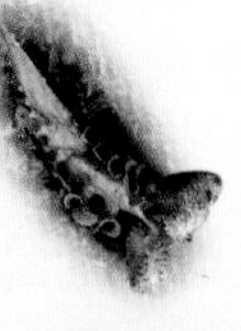

FURBELOWS *is a leathery, brown seaweed, with a large holdfast that can withstand wave action.*

Shingle

Driven into mounds by storm waves, shingle banks will support only those plants that are able to tolerate quick drainage and substrate mobility. Such plants often have succulent leaves and stems, cushion growth forms, and tap roots for anchorage.

SEA PEA *spreads in loose carpets over unstable shingle, and is anchored by deep roots.*

LITTLE TERN *breeds on bare shingle, its eggs camouflaged against the background.*

Sand

Wind-blown sand forms dunes, which are stabilized by drought-tolerant plants. Such plants often have silvery or hairy foliage, deep roots, waxy coatings, an annual life cycle, and a range of other adaptations to survive in a stressful environment.

SEA-HOLLY *has waxy, silvery leaves to reflect the sunlight, and thus help the plant to tolerate drought.*

GRAYLINGS *are camouflaged against a sandy background, when their wings are fully closed.*

Sand- and Mudflats

Tides bring both food and sediment onto the flats twice a day. This bounty attracts a large number of invertebrates, which in turn are a major draw for fish at high water, and birds at low tide when they can pick the invertebrates from the exposed mud and sand.

COMMON COCKLES *form dense beds, and are a major food source for wading birds.*

GUT-WEED *is tolerant of the low-salinity conditions found in estuarine mudflats, especially where nutrients drain off the land.*

Salt Marsh

All salt marshes are washed by tides – the lower marshes by every tide, but the upper marshes only by the highest tide. Plants in this environment require adaptations to cope with salt and waterlogging, such as succulence, air spaces in the roots, and salt excretion glands.

REDSHANKS *take advantage of lower summer tides to breed on high-level salt marshes.*

COMMON SEA-LAVENDER *is salt-tolerant, and spreads like a carpet of purple over marshes in mid-summer.*

Higher plants

The evolution of the seed (with its protective covering) allowed plants to colonize dry land, and there are relatively few aquatic species of higher (or seed-bearing) plant, especially in the salt-water environment. However, many plants are adapted to withstand salty, coastal environments. Some of the plants found on the coast also grow inland, but the species on the following pages are particularly associated with coastal habitats. The plants are primarily arranged by flower colour – yellow, white, pink and red, blue, purple, and green; the trees and shrubs feature at the end.

SEA WORMWOOD

THREE-HORNED STOCK

SEA CLUB-RUSH

AFRICAN TAMARISK

Yellow Horned-poppy

Glaucium flavum (Papaveraceae)

A colonist of bare shingle habitats, Yellow Horned-poppy is a very distinctive and colourful part of the flora of these areas. It is often found growing with Sea Kale (p.24) and Viper's-bugloss (p.43), producing a flowery mosaic of yellow, white, and blue in summer. Its waxy leaves, the lower ones with silvery hairs, help conserve water during summer droughts. It is a biennial or short-lived perennial, the greyish leaf rosettes persisting through the winter. A broken stem of the Yellow Horned-poppy exudes a characteristic yellowish latex.

OCCURS *on shingle banks and sandy beaches, occasionally sea-cliffs; disturbed areas further inland.*

NOTE

Like all poppies, the Yellow Horned-poppy has tiny seeds, which could get lost in shingle deposits as they are much smaller than the gaps between the pebbles. But this species has developed large fruit capsules, within which many of the seeds remain safely until germination.

SIZE *To 1m tall; flowers 6–9cm wide.*
FLOWERING TIME *April–September.*
LEAVES *Alternate, pinnately divided.*
FRUIT *Elongated capsule.*
DISTRIBUTION *Mediterranean; Atlantic, north to southern Norway.*
SIMILAR SPECIES Glaucium leiocarpum, *which has deeper yellow to brownish flowers, fruit less than 15cm long, and is found in eastern Mediterranean, can be found growing together with the Yellow Horned-poppy.*

Golden Samphire

Inula crithmoides (Asteraceae)

A woody-based perennial, Golden Samphire generally grows upright in the north of its range. It is often more bushy and sprawling around the Mediterranean, where it can dominate the arid banks of commercial salt pans. Its succulent leaves give it some tolerance of a salt-laden environment; when crushed, they exude a distinctive aroma of shoe polish.

FOUND *on higher level salt marshes, salt-sprayed cliffs, and around saline lagoons and salt pans.*

PERENNIAL

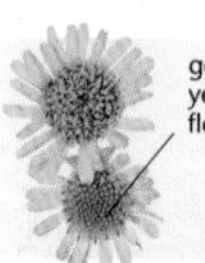

SIZE *Height to 1m; flowers minute, in heads 1.5–2.5cm wide.*
FLOWERING TIME *June–October.*
LEAVES *Linear, lance-shaped, yellowish green.*
FRUIT *Achene with a tuft of white hair.*
DISTRIBUTION *Mediterranean; Atlantic, north to southern Britain and Ireland.*
SIMILAR SPECIES *Sea Aster (p.42), the rayless form, which has broad, unscented leaves.*

Colt's-foot

Tussilago farfara (Asteraceae)

Spreading by rhizomes, Colt's-foot is able to colonize bare habitats quickly. Its flowers appear before the leaves, on stems clothed in overlapping scales. The broad, angled leaves expand to 60cm wide after flowering, and are often heavily colonized by a rust fungus.

OCCUPIES *soft, slumping cliffs, especially near water seepages; also widespread inland on road verges and damp, disturbed ground.*

PERENNIAL

SIZE *Height 15–25cm; flowers minute, in heads 1.5–3.5cm wide.*
FLOWERING TIME *February–April.*
LEAVES *Basal, horse hoof-shaped, downy, white and hairy underneath.*
FRUIT *Clock of feathered achenes.*
DISTRIBUTION *Throughout the region.*
SIMILAR SPECIES *Summer leaves could be mistaken for Butterbur* (Petasites *sp.*).

Cottonweed

Otanthus maritimus (Asteraceae)

With stems and leaves densely clothed in a white, woolly covering, Cottonweed is able to reflect the sun's rays and conserve water. This allows the plant to thrive in inhospitable, drought-prone habitats, where it forms extensive spreading patches. A southern species, it is now extinct in Britain, although one outlying population remains on the east coast of Ireland.

THRIVES *on sand dunes and stabilized shingle, often forming large patches on the seaward fringes.*

yellow flowerhead, lacking ray florets

leaves to 1cm long

PERENNIAL

SIZE *Height to 30cm; flowers minute, in heads 7–10mm wide.*
FLOWERING TIME *June–October.*
LEAVES *Alternate, oblong, toothed.*
FRUIT *Curved, ribbed achene.*
DISTRIBUTION *Mediterranean, Atlantic.*
SIMILAR SPECIES *Sea Wormwood (below);* Santolina chamaecyparissus, *which is taller and shrubby, with finely pinnate leaves.*

Sea Wormwood

Seriphidium maritimum (Asteraceae)

A low, woody-based perennial, Sea Wormwood is a conspicuously silvery plant that is pleasantly aromatic when its leaves are rubbed. Its flowers, while individually small, are borne in long drooping spikes, the narrow petals being a bright golden-yellow. Its growth form is variable, from erect to sprawling. Clumps of this plant often form a distinct silvery band on the upper salt marsh zone.

OCCURS *on the upper levels of salt marshes and on sea walls, where only the highest tides reach.*

PERENNIAL

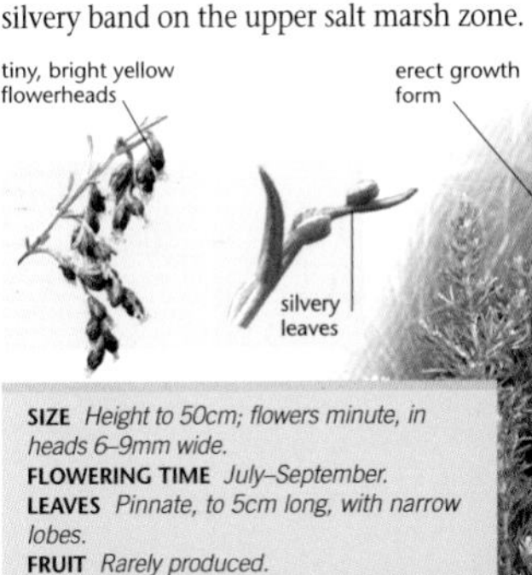

SIZE *Height to 50cm; flowers minute, in heads 6–9mm wide.*
FLOWERING TIME *July–September.*
LEAVES *Pinnate, to 5cm long, with narrow lobes.*
FRUIT *Rarely produced.*
DISTRIBUTION *Mediterranean; Atlantic, north to Scotland and southern Sweden.*
SIMILAR SPECIES *Cottonweed (above).*

Sea Medick

Medicago marina (Fabaceae)

FORMS *mats on dunes and upper sandy beaches, often in well-trampled areas near tourist resorts.*

The silvery, woolly coating of hair on Sea Medick gives the plant protection from drought and strong sunlight. It bears downy leaves in two regular ranks along the stem, and has small pea-flowers, which grow in clusters of 5–12. It can be easily distinguished from Strand Medick (*M. littoralis*), which has hairless pods with long spines, greener leaves, and is an annual.

PERENNIAL

SIZE *Height to 10cm; flowers 6–8mm long.*
FLOWERING TIME *February–June.*
LEAVES *Trifoliate, oval; downy leaflets.*
FRUIT *Spiralled pod with spines, 5–7mm long.*
DISTRIBUTION *Mediterranean; Atlantic, north to Brittany.*
SIMILAR SPECIES *Other* Medicago *species, which are less hairy and have distinctive seed pods.*

Southern Bird's-foot-trefoil

Lotus creticus (Fabaceae)

FOUND *on open sand dunes, and other coastal habitats such as low, rocky cliff-tops.*

Often replacing the Common Bird's-foot-trefoil (*L. corniculatus*) on Mediterranean dunes, Southern Bird's-foot-trefoil is distinguished by its often longer and more slender seed pods. These pods are splayed out like a bird's foot as with all members of this large genus. The leaves comprise five leaflets, each of which is broadest just above the middle.

PERENNIAL

SIZE *Height to 15cm; flowers 1.2–1.8cm long.*
FLOWERING TIME *March–May.*
LEAVES *Oblong, silvery, hairy leaflets.*
FRUIT *Straight or curved pod, to 5cm long.*
DISTRIBUTION *Mediterranean (excluding the islands) and Portugal.*
SIMILAR SPECIES L. cytisoides, *which has smaller flowers and a curved keel, and is widespread around the Mediterranean.*

Hairy Buttercup

Ranunculus sardous (Ranunculaceae)

This buttercup is easily recognizable, even from a distance, by its yellowish green leaves. Its five-petalled flowers are set within five green sepals, which reflex backwards to the stem at flowering time. Bulbous Buttercup (*R. bulbosus*) also shares this feature, but has deep yellow flowers and a swollen stem-base.

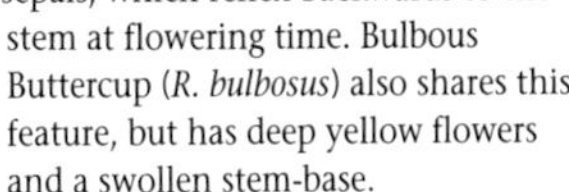

yellowish green leaf

lemon-yellow flower

OCCUPIES *seasonally flooded depressions in coastal grazing marshes; also found in low-lying grassland.*

SIZE *Height to 20cm; flowers 1.2–2cm wide.*
FLOWERING TIME *April–October.*
LEAVES *Three-lobed lower leaves, the central leaf larger and often lobed; hairy.*
FRUIT *Short-beaked achene, 2–3mm long.*
DISTRIBUTION *Mediterranean, Atlantic, southern Baltic.*
SIMILAR SPECIES R. bulbosus *(see above);* R. repens, *which has erect sepals.*

Silverweed

Potentilla anserina (Rosaceae)

A patch-forming perennial, Silverweed colonizes bare habitats by means of its rapidly growing red stolons, which run over the surface of the soil. The underside of the leaves is covered with silvery hairs, which contrast with the matt greyish green upper leaf surface, earning it the name Silverweed.

FAVOURS *the upper levels of sand and shingle beaches, and seepages in soft cliffs.*

5 bright yellow petals

SIZE *Height to 20cm; flowers 1.5–2cm wide.*
FLOWERING TIME *May–August.*
LEAVES *Basal rosettes; 10–20cm long, pinnate with 12–25 leaflets, each 2–5cm long.*
FRUIT *Tight head of dry achenes.*
DISTRIBUTION *Throughout the region, both coastal and inland.*
SIMILAR SPECIES *Other* Potentilla *species, which have palmate, not pinnate, leaves.*

Rock Samphire

Crithmum maritimum (Apiaceae)

OCCURS *on coastal rocks and cliffs; also sand and shingle banks.*

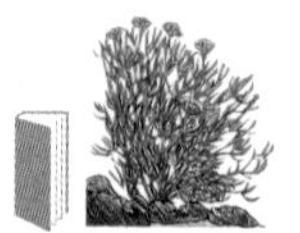

PERENNIAL

Like many seaside plants, Rock Samphire has fleshy, succulent leaves to conserve moisture in its arid, salt-laden environment. It can withstand salt spray but not regular immersion in salt water. Adopting different growth forms in different habitats, it usually scrambles on rocks and cliffs, but has a more cushion-like habit on sand and shingle banks. Despite its somewhat unappetizing smell of polish when crushed, it is frequently eaten, either raw, boiled, or pickled, especially in Mediterranean island cuisine. The tight, rounded umbels of the Rock Samphire are made up of small, creamy white flowers, which produce abundant nectar and attract many visiting insects.

NOTE

Also bearing the common name of Samphire are Golden Samphire (p.14) and Marsh Samphire (Glasswort, p.62). All are unrelated, linked only by their habitat and edibility. Similar confusion occurs in all languages, highlighting the need for standard scientific names.

SIZE *Height 20–50cm; umbels 3–6cm wide.*
FLOWERING TIME *May–August.*
LEAVES *Alternate, triangular, divided into cylindrical, fleshy, upward-pointing segments.*
FRUIT *Oval mericarp, ripening from yellow to purple; seeds often purplish when ripe, with prominent ridges.*
DISTRIBUTION *Mediterranean; Atlantic, north to Scotland.*
SIMILAR SPECIES *The yellow-flowered* Smyrnium olusatrum, *which also grows by the coast, but is upright, with broad leaflets.*

Dune Pansy

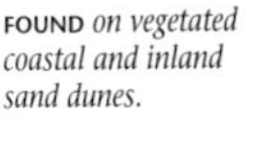

Viola tricolor ssp. *curtisii* (Violaceae)

A subspecies of the more familiar Heart's-ease (*V. tricolor tricolor*), Dune Pansy is distinguished by its perennial habit and spreading rhizomes. It also has wholly yellow flowers, alongside purple or bicoloured yellow and purple ones. The flat, five-petalled, almost face-like flowers have a spur projecting backwards, which produces nectar for visiting insects.

FOUND *on vegetated coastal and inland sand dunes.*

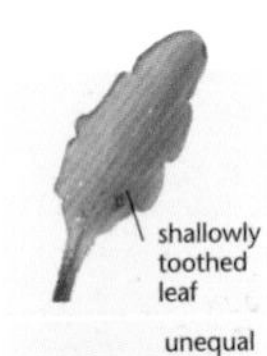

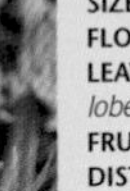

SIZE *Height to 15cm; flowers 1.5–2cm wide.*
FLOWERING TIME *April–September.*
LEAVES *Lower leaves ovate, with rounded lobes; upper leaves narrower.*
FRUIT *Capsule with three valves.*
DISTRIBUTION *Atlantic, south to northern France; Baltic.*
SIMILAR SPECIES V. kitaibeliana, *which has much smaller, creamy yellow flowers.*

Wild Cabbage

Brassica oleracea (Brassicaceae)

The wild precursor of many familiar vegetable greens, Wild Cabbage is a tall, hairless plant, with fleshy, lobed basal leaves and unlobed upper leaves, which clasp the stem. Thriving in nitrogen-rich habitats, it is often found around seabird colonies, its pale yellow flowers adding a splash of colour to the sensory mix of sounds and smells.

FAVOURS *sea cliffs and coastal grassland, especially on chalk and limestone.*

SIZE *Height to 1.2m; flowers 3–4cm wide.*
FLOWERING TIME *May–September.*
LEAVES *Wavy-edged, grey-green.*
FRUIT *Fleshy siliqua, 5–7cm long.*
DISTRIBUTION *Mediterranean; Atlantic, north to Britain.*
SIMILAR SPECIES Raphanus raphanistrum maritimum, *which has paler yellow flowers and swollen seed pods.*

Cistanche

Cistanche phelypaea (Orobanchaceae)

OCCUPIES *dry upper salt marsh habitats and the edges of Mediterranean salt pans, where its hosts are found.*

Totally lacking green pigmentation (chlorophyll), like its relatives, the broomrapes, Cistanche is a parasitic plant. It attaches its roots to that of the host plant from which it draws all of the food and water it needs, and so is restricted its host's habitat. The host plants are shrubby members of the spinach family (*Chenopodiaceae*), such as Shrubby Sea-blite (p.61) and Shrubby Glasswort (p.63).

waxy appearance

leaves reduced to scales on stems

flowers in dense spikes

tubular, shiny yellow flower

5 almost equal lobes

PERENNIAL

SIZE *Height to 60cm; flowers 3–4cm long.*
FLOWERING TIME *March–May.*
LEAVES *Alternate, scale-like, without chlorophyll.*
FRUIT *Capsule with numerous tiny seeds, splitting lengthwise.*
DISTRIBUTION *Mediterranean, Atlantic (Portugal).*
SIMILAR SPECIES *None.*

Biting Stonecrop

Sedum acre (Crassulaceae)

FORMS *extensive, low patches on shingle, stabilized sand, and cliffs; also inland.*

Appropriately named Biting Stonecrop or Wall Pepper, this plant has leaves with a burning, peppery taste. It adapts admirably to arid environments with its succulent leaves, which store water in dry conditions; its patch-forming habit, which conserves moisture; and the cylindrical, waxy-coated leaves, which minimize loss of water vapour.

PERENNIAL

SIZE *Height to 10cm; flowers 0.9–1.1cm wide.*
FLOWERING TIME *April–July.*
LEAVES *Alternate, succulent, 3–6mm long.*
FRUIT *Five follicles in star shape, to 4mm long.*
DISTRIBUTION *Western Mediterranean, Atlantic, Baltic.*
SIMILAR SPECIES S. sexangulare, *which has longer leaves, without the peppery taste.*

White Stonecrop

Sedum album (Crassulaceae)

Sometimes confused with the similarly coloured English Stonecrop (below), White Stonecrop is easily distinguished by its domed or flat flowerheads raised on a single stalk from its mat of fleshy leaves. Although widespread inland, it is often a conspicuous component of the coastal flora.

OCCURS *in a range of sand, shingle, and cliff habitats, above the reach of salt water and spray; similar habitats inland and walls.*

SIZE *Height to 15cm; flowers 6–9 mm wide.*
FLOWERING TIME *June–August.*
LEAVES *Alternate, small, cylindrical, and succulent.*
FRUIT *Cluster of small follicles.*
DISTRIBUTION *Mediterranean, Atlantic, southern Baltic.*
SIMILAR SPECIES *English Stonecrop (below), which does not have flowerheads.*

English Stonecrop

Sedum anglicum (Crassulaceae)

A mat-forming plant, English Stonecrop has a distinctive overall colour. Its white flowers and greyish leaves are often tinged with pink, especially in full sunlight. It is believed that the red pigment acts as a sunscreen, filtering the harmful rays of the sun. Like most *Sedums*, English Stonecrop is very succulent, with fleshy leaves – a further adaptation to its hot and dry habitat.

FAVOURS *shingle, sand, and cliff-top habitats, usually not on lime soil; similar habitats and walls inland.*

star-shaped flower

SIZE *Height to 5cm; flowers 1.1–1.2cm wide.*
FLOWERING TIME *June–September.*
LEAVES *Alternate, oval; clasping the stem.*
FRUIT *Cluster of carpels, erect, becoming red when ripe.*
DISTRIBUTION *Atlantic.*
SIMILAR SPECIES *White Stonecrop (above), which has flowerheads and lacks any reddish tinge on the leaves.*

Common Scurvy-grass

Cochlearia officinalis (Brassicaceae)

OCCURS *in the upper reaches of salt marshes, and on cliffs and sea walls.*

Neither a grass nor particularly rich in Vitamin C, which would prevent scurvy, the inappropriately named Common Scurvy-grass is actually a member of the cabbage family. It has white flowers with four rounded petals. The leaves are succulent, the basal leaves stalked, the upper leaves clasping the stem.

BIENNIAL/PERENNIAL

SIZE *Height to 50cm; flowers 8–10mm wide.*
FLOWERING TIME *April–August.*
LEAVES *Mostly basal, rounded, and fleshy; upper leaves are stalkless.*
FRUIT *Swollen silicula, with cork-like texture.*
DISTRIBUTION *Atlantic, southern Baltic.*
SIMILAR SPECIES C. anglica, *which has lower leaves tapering at the base, compressed fruit, and is restricted to salt marshes.*

Early Scurvy-grass

Cochlearia danica (Brassicaceae)

OCCUPIES *the transition zone between salt marsh and sand dunes; bare cliff-tops and the verges of inland roads.*

A diminutive plant, Early Scurvy-grass flowers very early in the year. It bears small flowers with petals that are less than twice as long as the sepals, and ivy-like, lobed upper leaves. Recently, it has spread extensively inland along major roads, especially on the central reservations, where salting of roads in winter provides conditions that are similar to its natural habitats.

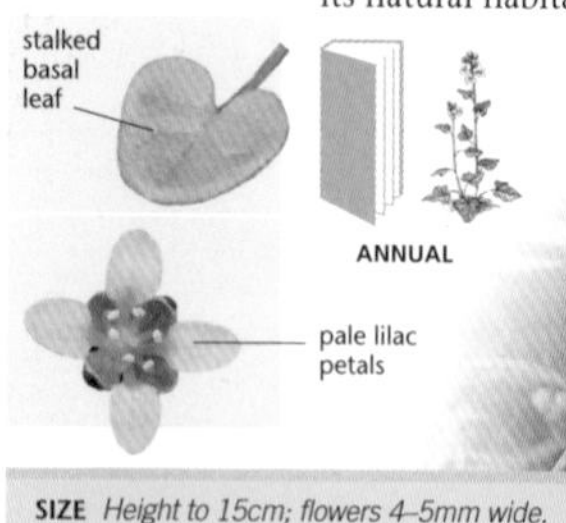

SIZE *Height to 15cm; flowers 4–5mm wide.*
FLOWERING TIME *January–July.*
LEAVES *Basal leaves long-stalked, rounded; stem leaves deeply lobed, often unstalked.*
FRUIT *Broadly oval seed pod, with small reddish brown seeds, 3–6mm long.*
DISTRIBUTION *Atlantic, Baltic.*
SIMILAR SPECIES *Other* Cochlearia *species, such as Common Scurvy-grass (above).*

Sea Rocket

Cakile maritima (Brassicaceae)

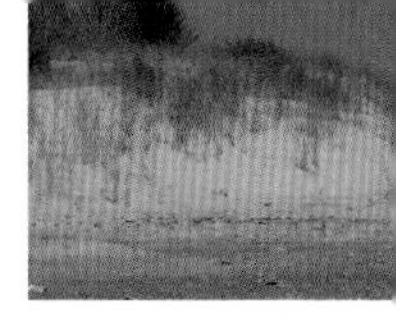

GROWS *on sandy drift lines, extending upwards onto the embryonic parts of sand dune systems.*

Sea Rocket almost invariably grows at the highest tide mark of sandy coastlines. It has two types of seeds: large, corky seeds, which are deposited on a drift line, where they may or may not germinate the following spring, and smaller ones that remain attached to the plant, and so have a more guaranteed germination site.

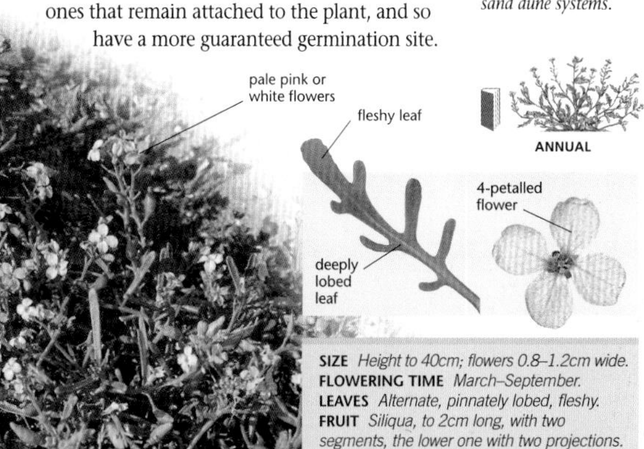

SIZE *Height to 40cm; flowers 0.8–1.2cm wide.*
FLOWERING TIME *March–September.*
LEAVES *Alternate, pinnately lobed, fleshy.*
FRUIT *Siliqua, to 2cm long, with two segments, the lower one with two projections.*
DISTRIBUTION *Throughout the region.*
SIMILAR SPECIES *Three-horned Stock (p.35) and* Matthiola sinuata, *which have longer pods and larger flowers.*

Early Whitlow-grass

Erophila verna (Brassicaceae)

FOUND *on sand dunes and other bare, sandy habitats; often conspicuous on ant hills in grazing marshes.*

A tiny early flowering species, Early Whitlow-grass is a classic winter annual: it germinates in autumn, then flowers, sets seed, and dies before the onset of dry conditions, surviving as drought-tolerant seeds. It bears white flowers in loose spikes on leafless stems, arising from a rosette of leaves to 6cm wide. Each flower petal is divided almost to the base into two lobes.

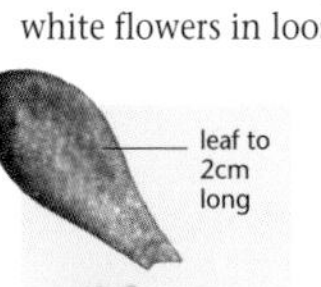

leafless flowering stem

4-petals, each 2-lobed

SIZE *Height to 15cm; flowers 3–5mm wide.*
FLOWERING TIME *January–April.*
LEAVES *Stalked, 1 to 2 teeth on either side; hairy.*
FRUIT *Flattened, oval seed pod with up to 50 small seeds; 1.5–3 times longer than its width.*
DISTRIBUTION *Throughout the region.*
SIMILAR SPECIES *Several other* Erophila *species, which differ in pod and petal shape.*

Sea Kale

Crambe maritima (Brassicaceae)

FORMS *large clumps on otherwise almost bare shingle banks, or less frequently on upper sandy beaches and low clay cliffs.*

The large clumps of waxy, lobed, cabbage-like leaves of Sea Kale are one of the most distinctive sights on shingle banks and spits in northwestern Europe. In mid-summer, a plant that is large enough to flower produces huge domed heads of white flowers. These are later replaced by heads of rounded seed pods, each the size and shape of a marble, and bearing a single seed. The pods float and are dispersed by the sea, while the parent plant remains anchored in the shingle by means of its deep tap root.

PERENNIAL

4-petalled flowers in large clusters

NOTE

The leaves and stems of Sea Kale were formerly blanched for eating by piling shingle or beach debris upon the emerging shoots in spring.

SIZE *Height to 1m; spread to 2m; flowers 1–2cm wide.*
FLOWERING TIME *June–August.*
LEAVES *Mostly basal, lobed with wavy margins, grey-green; lower leaves unstalked.*
FRUIT *Fleshy, rounded silicula, 0.8–1.4cm long.*
DISTRIBUTION *Atlantic, from northern Spain to south Norway; southern Baltic.*
SIMILAR SPECIES Armoracia rusticana, *which has unlobed, wavy-edged, erect leaves, and smaller flowers and fruit.*

Sea Sandwort

Honckenya peploides (Caryophyllaceae)

A distinctive upper beach plant, Sea Sandwort spreads easily with its extensive spaghetti-like stolon system. It bears shiny, fleshy leaves in pairs. It has separate male and female flowers, but the yellow-green seed capsules are more likely to attract attention than these.

GROWS *on seaward sand dunes and upper beaches, forming fore-dunes; also on shingle.*

SIZE *Height to 20cm; male flowers to 1cm across, female flowers to 0.6cm across. 7–15cm long; flower lip 1.1–1.3cm long.*
FLOWERING TIME *May–August.*
LEAVES *Opposite pairs, at right angles to the pair above and below, oval, and stalkless.*
FRUIT *Many-seeded, rounded capsule.*
DISTRIBUTION *Atlantic, Baltic.*
SIMILAR SPECIES *None.*

Sea Mouse-ear

Cerastium diffusum (Caryophyllaceae)

A small, autumn-germinating plant of sand dunes, the stems and leaves of Sea Mouse-ear are clothed in sticky, glandular hairs, which often trap grains of sand. Unlike most members of its genus, its flower parts – sepals, petals, stamens, and styles – are usually in fours, although the occasional five-parted specimens also occur.

FOUND *on sand dunes and shingle banks; also on rocks and cliffs, and on dry heaths inland.*

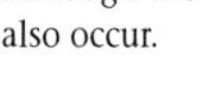

white petals, deeply divided into 2 lobes

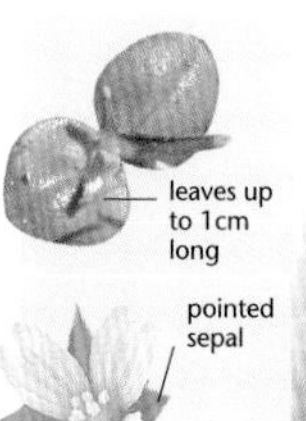

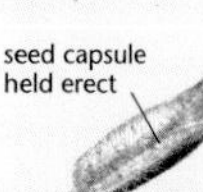

SIZE *Height to 10cm; flowers 3–6mm wide.*
FLOWERING TIME *March–August.*
LEAVES *Ovate, with sticky glandular hairs.*
FRUIT *Narrowly cylindrical capsule, to 8mm long, on stalks much longer than the capsule.*
DISTRIBUTION *Mediterranean, Atlantic, southern Baltic*
SIMILAR SPECIES C. semidecandrum, *which has 5-parted flowers and recurved fruit stalks.*

Sea Campion

Silene uniflora (Caryophyllaceae)

OCCURS *in contrasting coastal habitats: shingle banks and cliffs; inland on high mountains.*

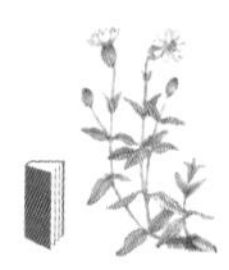

Varying its growth form according to its habitat, Sea Campion is a loose, scrambling plant on rocky cliffs and grassy cliff slopes. In contrast, it forms tight, rounded cushions on shingle in order to avoid wind damage, as well as to conserve moisture within its dense, matted foliage. Some of its waxy, hairless leaves remain green throughout the winter, which distinguishes it from a number of related inland species. The large flowers are usually borne singly or in small groups at the tips of the shoots. However, it also has numerous non-flowering shoots.

SIZE *Height to 50cm; flowers to 2–2.5cm wide.*
FLOWERING TIME *March–October.*
LEAVES *Opposite, oval, untoothed; only the lowermost leaves are stalked.*
FRUIT *Many-seeded capsule with six teeth.*
DISTRIBUTION *Western Mediterranean, Atlantic, Baltic.*
SIMILAR SPECIES S. nutans, *which is also a shingle plant, but has recurved petals and drooping flower spikes;* S. vulgaris, *which has even more inflated sepal tubes.*

NOTE

Sea Campion was earlier considered a subspecies of the Bladder Campion (S. vulgaris), *which may be found in cliff-top grassland but withers in the winter.*

Marsh Helleborine

Epipactis palustris (Orchidaceae)

With its long creeping rhizomes, this orchid often forms dense colonies in suitable damp habitats. Its leaves, which are otherwise difficult to distinguish from those of many other orchids, are often purplish below. The flowers are a combination of white, pink, yellow, and green, often with a reddish tinge that varies between plants.

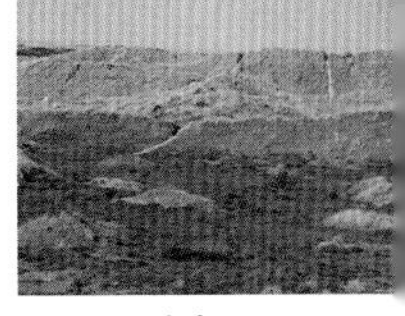

OCCUPIES *salt-free areas, especially dune slacks and soft cliff spring lines; fens and marshes inland.*

PERENNIAL

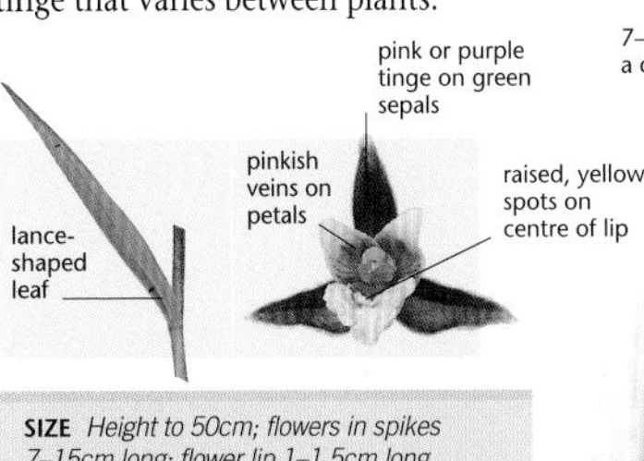

SIZE *Height to 50cm; flowers in spikes 7–15cm long; flower lip 1–1.5cm long.*
FLOWERING TIME *June–August.*
LEAVES *Alternate, lance-shaped.*
FRUIT *Three-parted, pendent capsule.*
DISTRIBUTION *Throughout the region; scarce in the extreme north and south.*
SIMILAR SPECIES *Several other sand dune helleborines, which have green flowers.*

Sea Daffodil

Pancratium maritimum (Liliaceae)

An attractive plant of Mediterranean beaches, the Sea Daffodil is now less frequently seen as development encroaches upon its habitat. The tubular, brilliant white flowers, several on each spike, are strongly and sweetly scented and attract long-tongued insects such as Convolvulus Hawk-moths. The leaves, which emerge in winter, usually wither by flowering time.

FOUND *on sand dunes and upper sandy beaches, often in close proximity to resorts.*

PERENNIAL

strap-like, twisted leaf

black seeds

SIZE *Height to 50cm; flowers 10–15cm long.*
FLOWERING TIME *July–October.*
LEAVES *Narrow, to 2cm wide, blue-green.*
FRUIT *Large, three-sided, rounded capsule, with triangular black seeds.*
DISTRIBUTION *Mediterranean; Atlantic, north to southern Brittany.*
SIMILAR SPECIES P. illyrica, *which has a shorter corona and wider leaves.*

Sea Knotgrass

Polygonum maritimum (Polygonaceae)

FAVOURS *upper sandy and shingle beaches, and the seaward fringe of sand dunes; is a recent recolonist of southern Britain.*

A low, sprawling plant with a woody base, Sea Knotgrass forms loose, bluish grey patches on upper beaches, especially in southern Europe. It is relatively tolerant of trampling. The seeds are dispersed by sea currents, but the plant's habitat is vulnerable to erosion by the sea.

PERENNIAL

SIZE *Height to 15cm; spread 0.5m–1m; flowers 2–4mm wide.*
FLOWERING TIME *July–September.*
LEAVES *Alternate, lance-shaped.*
FRUIT *Reddish brown nutlet, to 2.5mm long.*
DISTRIBUTION *Mediterranean; Atlantic, as far north as southern Britain.*
SIMILAR SPECIES P. oxyspermum, *which has larger, shiny, dark brown nutlets.*

Round-leaved Wintergreen

Pyrola rotundifolia (Pyrolaceae)

GROWS *in dune slacks, often with Creeping Willow; inland in bogs and damp, shady woods.*

Usually referred to as subspecies *maritima*, the dune slack form of Round-leaved Wintergreen differs from its inland counterpart in only minor, inconsistent features. Its white flowers are borne in a spike of up to 10 flowers, each with a single protruding, curved style. Its rosettes of glossy rounded leaves are distinctive throughout the year.

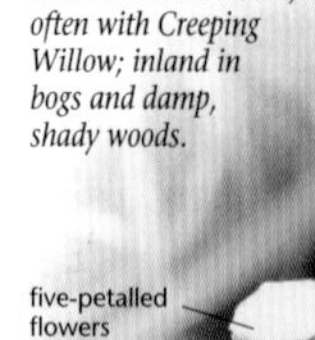

PERENNIAL

SIZE *Height to 20cm; flowers 0.8–1.2cm wide.*
FLOWERING TIME *June–September.*
LEAVES *Basal, shallowly toothed; dark green and glossy, to 5cm long.*
FRUIT *Capsule with many small seeds.*
DISTRIBUTION *Atlantic, from northwest France to Germany and northern Britain.*
SIMILAR SPECIES Parnassia palustris, *which has single, upright, but similar-looking flowers.*

Echinophora

Echinophora spinosa (Apiaceae)

Sometimes called Sea Parsnip or Prickly Samphire, this plant is edible, although the leaves develop a tough coating and spiny tips soon after emergence. In flower, it forms distinctive pale clumps, often growing with Sea Holly (p.45) and Sea Daffodil (p.27). The partial umbels consist of several male flowers surrounding a single hermaphrodite flower.

FOUND *towards the seaward edge of relatively unvegetated Mediterranean sand dune systems.*

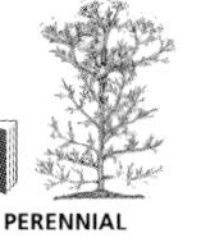

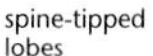

SIZE *Height to 60cm; small flowers in umbels to 10cm wide.*
FLOWERING TIME *June–October.*
LEAVES *Pinnate, with grooved lobes.*
FRUIT *Egg-shaped, with indistinct ridges.*
DISTRIBUTION *West and central Mediterranean.*
SIMILAR SPECIES *Sea Holly (p.45);* E. tenuifolium, *which has yellow flowers.*

Scots Lovage

Ligusticum scoticum (Apiaceae)

The shiny green Scots Lovage, with its stiff, erect stems and greenish white umbels of small flowers, is a characteristic plant of more northerly rocky coastlines. Its leaves are described as twice-ternate – they are divided into three almost equal lobes, which are further divided into three broad leaflets. These are toothed, but only along the upper half of the leaflet.

INHABITS *low cliffs and rocky places near the sea in the north; locally around the Baltic.*

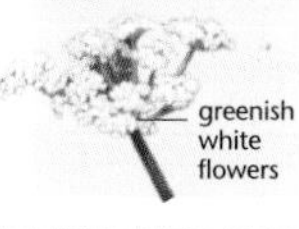

SIZE *Height to 90cm; umbels 4–6cm wide.*
FLOWERING TIME *June–July.*
LEAVES *Divided into broad, ovate, leathery lobes, to 5cm long; toothed towards the tip; stalks inflated, sheathing the stem.*
FRUIT *Two fused carpels, sharply ridged.*
DISTRIBUTION *Atlantic, north from Scotland; Baltic.*
SIMILAR SPECIES *None.*

Sea Mayweed

Tripleurospermum maritimum (Asteraceae)

OCCURS *on drift lines, sand dunes, and rocks and cliffs, especially around seabird colonies.*

Although it can form more dense cushions, especially on shingle, Sea Mayweed is often a loose, trailing plant. The succulent leaves help the plant to withstand the dry, salty conditions of its habitat. It favours nitrogen-rich locations, such as bird colonies and old drift lines of rotting seaweed.

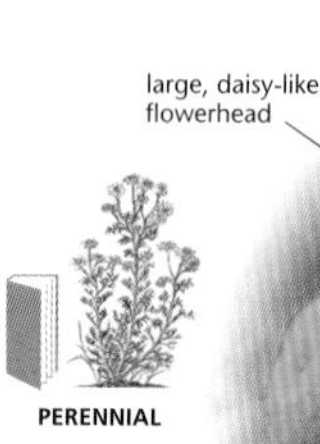

SIZE *Height to 50cm; flowerhead 3–5cm wide.*
FLOWERING TIME *July–September.*
LEAVES *Alternate, finely divided.*
FRUIT *Ridged, blackish brown achene, 3–5mm long.*
DISTRIBUTION *Atlantic and Baltic.*
SIMILAR SPECIES *Several similar species are primarily cornfield annuals, which have smaller flowers, and leaves that are not fleshy.*

Ice Plant

Mesembryanthemum crystallinum (Aizoaceae)

THRIVES *in extremely salty areas of sand dunes, salt marshes, and salt pans.*

One of the few native members of this family of succulents, most of which originate from South Africa, Ice Plant is named for the sparkling, crystalline appearance of its leaves, covered in glistening hairs. Its flowers are solitary, with numerous linear, pale yellow or whitish petals that are much longer than the sepals.

SIZE *Height to 5cm; flowers 2–3cm wide.*
FLOWERING TIME *May–August.*
LEAVES *Alternate, pointed, oval to spoon-shaped, untoothed.*
FRUIT *Small, fleshy capsule.*
DISTRIBUTION *Mediterranean.*
SIMILAR SPECIES M. nodiflorum, *which has petals shorter than the sepals, and narrow leaves, and is also found in the Mediterranean.*

Hottentot Fig

Carpobrotus edulis (Aizoaceae)

A native of South Africa, Hottentot Fig is widely naturalized in southern coastal areas of Europe. Its invasive nature means that much effort goes into preventing it from spreading – this plant can regrow rapidly from small stem and leaf fragments. The flowers have two colour forms, magenta and yellow, which often grow together, carpeting large areas of cliffs and coastal rocks. The leaves are succulent and toothed on the keel. Although edible, the fig-like fruit is neither juicy nor tasty.

FORMS *a low, creeping carpet on sandy ground and cliff slopes; often cultivated and frequently escaping from gardens.*

SIZE *Height to 25cm; flowers 7–9cm wide.*
FLOWERING TIME *March–July.*
LEAVES *Opposite, fleshy and triangular in cross-section, widest at the base; joined around the stem at their base.*
FRUIT *Fleshy, many-seeded capsule.*
DISTRIBUTION *Mediterranean; Atlantic, north to Ireland.*
SIMILAR SPECIES C. acinaciformis, *which has larger, deeper pink flowers with purple stamens, and leaves that are broadest at or above the middle.*

NOTE

An increasing weed of sand dunes and cliff slopes, Hottentot Fig is spreading at the expense of native plants. Although it is sensitive to frosts, its range is still expanding, probably as a result of climate change.

Marsh Mallow

Althaea officinalis (Malvaceae)

A beautiful plant with very pale pink flowers and downy, greyish leaves, Marsh Mallow is often found among Common Reeds (p.52) in coastal waters and marshland. The root, with its high mucilage content, was formerly used to make marshmallow sweets.

FAVOURS *upper salt marshes, brackish ditches, and tidal river banks; also extensive inland marshes.*

PERENNIAL

coarsely toothed leaf

SIZE *Height to 1.5m; flowers 2.5–4cm wide.*
FLOWERING TIME *July–September.*
LEAVES *Alternate, triangular, with 3–5 lobes and long-stalked; softly hairy.*
FRUIT *Hairy mericarps in a ring, surrounded by the calyx.*
DISTRIBUTION *Mediterranean; Atlantic, north to Denmark.*
SIMILAR SPECIES *None.*

Tree-mallow

Lavatera arborea (Malvaceae)

The stately, robust Tree-mallow is often grown as a garden plant, away from the risk of frosts. As a result, its native distribution is partly obscured by escapes into the wild. It has a stout woody stem and palmately lobed leaves. The cup-shaped flowers have deep magenta-pink, broadly overlapping petals, with dark lines radiating out from a blackish centre.

OCCURS *on cliff slopes, shingle, dunes, and open areas; perhaps naturalized in northerly parts of its range.*

BIENNIAL

SIZE *Height to 3m; flowers 3–4cm wide.*
FLOWERING TIME *June–September.*
LEAVES *Alternate, pale below when young.*
FRUIT *Cluster of nutlets in a ring.*
DISTRIBUTION *Mediterranean; Atlantic, north to Britain and Ireland.*
SIMILAR SPECIES L. cretica, *which is shorter, with deeply notched petals;* Malva sylvestris, *which has paler flowers.*

Hound's-tongue

FOUND *in grazed sand dune turf; inland on dry, rough, and chalky grassland sites.*

Cynoglossum officinale (Boraginaceae)

Exuding a characteristic odour of mouse urine that deters animals, Hound's-tongue remains untouched when all other plants around it are grazed. A roughly hairy plant, it has branched flowering stems, with flower spikes elongating in fruit. The tubular flowers, consisting of five fused petals, are followed by nutlets, covered in hooked hairs to facilitate dispersal.

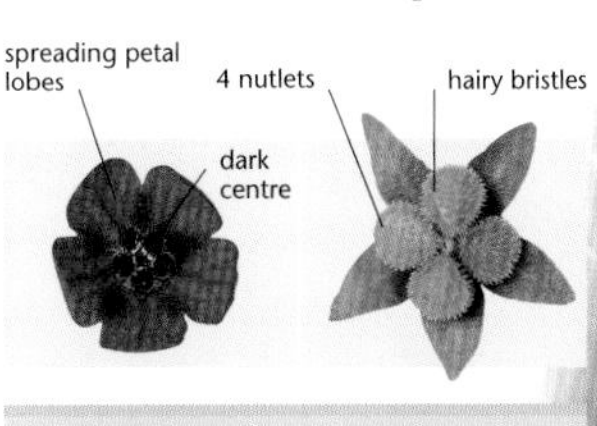

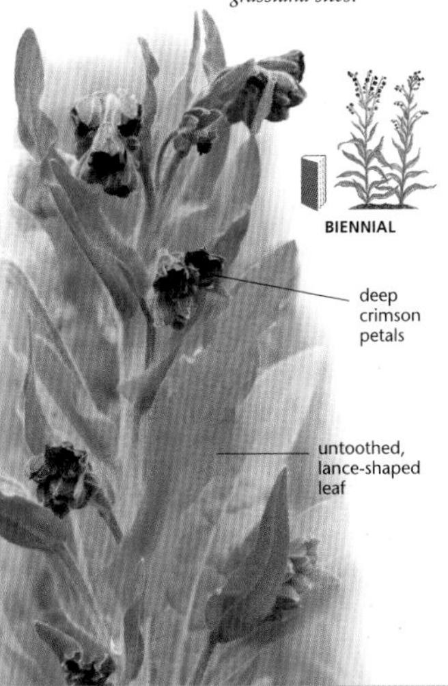

SIZE *Height to 90cm; flowers 6–10mm wide.*
FLOWERING TIME *May–August.*
LEAVES *Alternate, basal, coarsely hairy.*
FRUIT *Four nutlets, with a thickened border.*
DISTRIBUTION *Mediterranean, Atlantic, southern Baltic.*
SIMILAR SPECIES C. creticum, *which has blue flowers with purplish veins;* C. cheirifolium, *which has purple flowers.*

Early Marsh-orchid

FAVOURS *dune slacks and soft cliff seepages; inland forms occur in marshes, bogs, and other wet areas.*

Dactylorhiza incarnata (Orchidaceae)

A taxonomically difficult and variable species, Early Marsh-orchid occurs in a range of forms, which differ in flower colour and other features. The mainly coastal subspecies, *D. i. coccinea*, normally has brick-red flowers and broader leaves than inland forms. The lower lip of the flower appears narrow, on account of the sides of the lip, which are usually strongly curved back towards the stem.

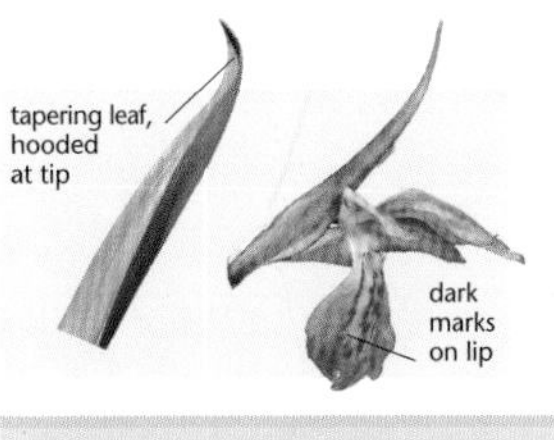

SIZE *Height to 40cm; flower lip to 9mm long.*
FLOWERING TIME *May–August.*
LEAVES *Basal rosette and spirally arranged up the stem; usually unspotted.*
FRUIT *Capsule with numerous, dust-like seeds.*
DISTRIBUTION *Mediterranean (rare); Atlantic, north to southern Norway; southern Baltic.*
SIMILAR SPECIES *Other marsh-orchids, although few have the apparently narrow lip.*

Strawberry Clover

Trifolium fragiferum (Fabaceae)

OCCURS *in damp meadows and coastal grazing marshes; scattered inland in damp grassland on clay.*

A creeping, patch-forming plant that roots at its nodes, Strawberry Clover is named after its reddish brown, strawberry-like fruiting heads. These heads comprise many inflated calyces, each surrounding a small seed pod. The delicate, pale pink flowers are relatively inconspicuous, and are borne in flowerheads 1–2cm wide.

SIZE *Height to 15cm; flowers 6–7mm long.*
FLOWERING TIME *June–September.*
LEAVES *Alternate, with oval leaflets.*
FRUIT *Head of downy calyces.*
DISTRIBUTION *Mediterranean, Atlantic, southern Baltic.*
SIMILAR SPECIES T. repens, *which has longer flowers;* T. resupinatum, *which has upside-down, splayed flowers.*

Sea Clover

Trifolium squamosum (Fabaceae)

FAVOURS *short, often grazed, turf in grassland close to the sea; abundant on grassy sea walls.*

A short, erect plant of coastal grassland, Sea Clover has rounded, pale pink flowerheads, which usually have a pair of leaves just below them. Each leaf consists of three leaflets, which are very narrow and pointed. When in fruit, the flowerhead elongates into an egg shape, and the starry, spreading sepals give it a very distinctive appearance.

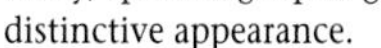

SIZE *Height to 40cm; flowers 7–9mm long.*
FLOWERING TIME *June–August.*
LEAVES *Mostly alternate, trifoliate.*
FRUIT *Small pod within spreading sepals.*
DISTRIBUTION *Mediterranean; Atlantic, north from Britain.*
SIMILAR SPECIES T. stellatum, *which has longer sepals in fruit, flowers maturing from pink to white, and rounded leaflets.*

Thrift

Armeria maritima (Plumbaginaceae)

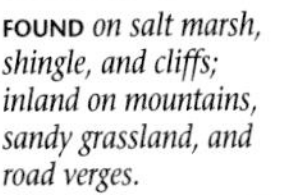

FOUND *on salt marsh, shingle, and cliffs; inland on mountains, sandy grassland, and road verges.*

Also known as Sea Pink, Thrift creates a dramatic splash of early summer colour in many coastal habitats. Its numerous flowerheads grow in abundance from cushions of fleshy, evergreen leaves. The flower colour is variable, from almost white to a deep purplish pink, the latter being the most frequent form grown in gardens.

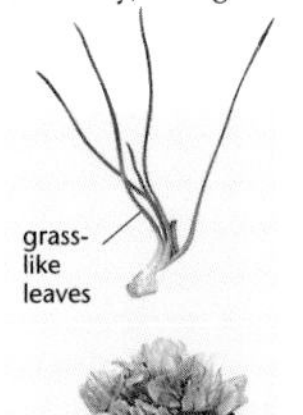

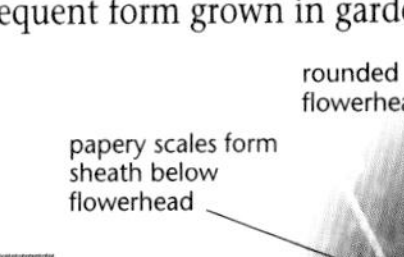

SIZE *Height to 30cm; flowers 8mm wide, in flowerheads 1.5–3cm wide.*
FLOWERING TIME *April–October.*
LEAVES *Basal, linear.*
FRUIT *Small, one-seeded capsule, with a papery wall.*
DISTRIBUTION *Atlantic, southern Baltic.*
SIMILAR SPECIES *Other* Armeria *species, which differ in size, colour, and leaf shape.*

Three-horned Stock

Matthiola tricuspidata (Brassicaceae)

OCCUPIES *coastal sandy areas, especially upper beaches and fore-dunes.*

Three-horned Stock is a colonist of bare sandy areas. Somewhat variable, its leaves are covered in loose, woolly hairs, are oblong in outline, unstalked, and shallowly lobed to pinnate. The four-petalled flowers appear in loose clusters, their colour ranging from lilac to purple. The most distinctive feature of the plant is the three-pointed apex to the long seed pods.

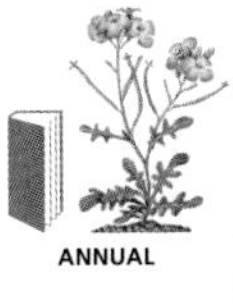

SIZE *Height to 40cm; flowers 1.8–2.5cm wide.*
FLOWERING TIME *February–May.*
LEAVES *Lobed to pinnate.*
FRUIT *Pod, 5–10cm long.*
DISTRIBUTION *Mediterranean.*
SIMILAR SPECIES *Sea Rocket (p.23);* M. incana, *which is perennial and has one point on pod;* M. sinuata, *which has toothed leaves.*

Sea Bindweed

Calystegia soldanella (Convolvulaceae)

CRAWLS *over sand dunes, upper sandy shores, and coastal shingle, often within other vegetation.*

NOTE

Its extensively creeping rhizomes and stems allow Sea Bindweed to seek out water in its dry habitats. The succulent leaves act as a reservoir for times of need.

A creeping plant, Sea Bindweed grows on stems that are about a metre or more long. It differs from many of its relatives in that it rarely twines around neighbouring plants to add to its height. It has fleshy, kidney-shaped leaves, which are very different from the pointed, often flimsy leaves of other bindweeds. Its large, trumpet-shaped flowers are pink, with a yellowish centre and five radiating white stripes. Surrounding the sepals are two papery bracts or scales – a feature that distinguishes the *Calystegia* species from bindweeds of the related genus *Convolvulus*. The flexible stems are quite strong and are sometimes used as makeshift string.

SIZE *Height to 30cm; flowers 3–5cm long and up to 4cm wide.*
FLOWERING TIME *May–August.*
LEAVES *Alternate; stalks longer than the blade.*
FRUIT *Pointed, oval, many-seeded capsule.*
DISTRIBUTION *Mediterranean; Atlantic, north to Denmark.*
SIMILAR SPECIES Convolvulus arvensis, *a frequent weed, which has smaller flowers and pointed leaves;* Ipomoea sagittata, *which has arrow-shaped leaves and pink flowers, and is found in Mediterranean salt marshes.*

Bog Pimpernel

Anagallis tenella (Primulaceae)

FAVOURS *damp, seasonally flooded dune slacks; inland in bogs and other peaty habitats.*

Bog Pimpernel has pink flowers that open fully only in bright sunlight. These flowers have slender stalks that arise from the leaf axils. The leaves, each borne on a short stalk, form a loose, straggling carpet on the ground. Each leaf has small, brown spots, or oil glands, on the underside margin. When crushed, they emit a surprisingly antiseptic smell. This plant creeps over dune slacks, although only on sand with a low lime content.

PERENNIAL

oval leaf

funnel-shaped flower

SIZE *Height to 10cm; flowers 0.6–1.2cm wide.*
FLOWERING TIME *June–August.*
LEAVES *Opposite and entire (unnotched, with smooth margin); each leaf distinctly stalked.*
FRUIT *Small rounded capsule, to 2mm wide.*
DISTRIBUTION *Atlantic, Baltic.*
SIMILAR SPECIES *Sea Milkwort (p.38), which has unstalked, fleshy leaves;* Montia fontana, *which has similar but unscented leaves.*

Scarlet Pimpernel

Anagallis arvensis (Primulaceae)

THRIVES *in naturally disturbed sand dune and shingle habitats. Widespread in disturbed areas inland.*

A familiar arable and garden weed, Scarlet Pimpernel also goes by the folk name of Poor Man's Weatherglass. This name is derived from the fact that its flowers open only in sunlight. The plant usually bears red flowers in northern Europe, while blue-flowered forms predominate in the south.

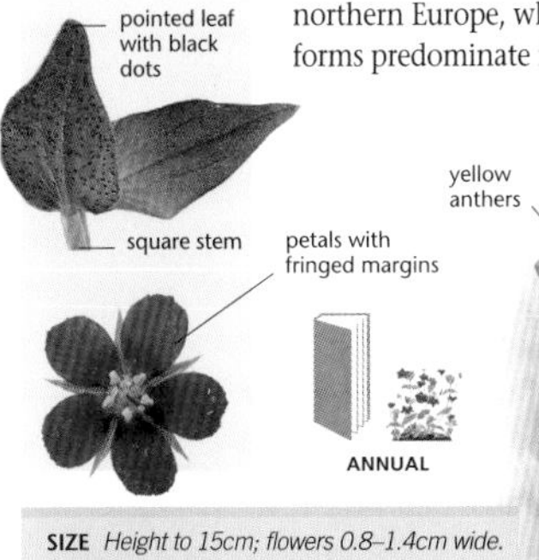

ANNUAL

SIZE *Height to 15cm; flowers 0.8–1.4cm wide.*
FLOWERING TIME *March–September.*
LEAVES *Opposite, unstalked, oval.*
FRUIT *Rounded capsule, 4–6mm wide.*
DISTRIBUTION *Mediterranean, Atlantic, southern Baltic.*
SIMILAR SPECIES A. foemina, *which has narrower blue petals, and* A. monelli, *which has linear leaves; both usually blue.*

Sea-heath

Frankenia laevis (Frankeniaceae)

FAVOURS *upper salt marsh habitats, especially on the transition zone with dunes; also shingle, coastal rocks, and cliffs.*

A mat-forming plant, Sea-heath can be distinguished from unrelated heathers by its flat, five-petalled flowers. The flowers are borne singly or in small clusters at the base of the leaves. En masse, the flowers seem sprinkled on the foliage mat, offering a striking sight. It is usually found within the reach of salt water or spray drift.

leaf to 5mm wide

leaves crowded on short side-shoots

dense, carpeting growth

small pink flower

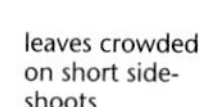

PERENNIAL

SIZE *Height to 15cm; flowers 0.8–1.2cm wide.*
FLOWERING TIME *June–August.*
LEAVES *Opposite, linear, with recurved margins.*
FRUIT *Conical capsule, concealed in the calyx.*
DISTRIBUTION *Western Mediterranean; Atlantic, north to Britain.*
SIMILAR SPECIES *Several other* Frankenia *species in the Mediterranean region, which have slightly different flowers and leaves.*

Sea Milkwort

Glaux maritima (Primulaceae)

OCCURS *in damp saline and brackish habitats, such as upper salt marshes, dune slacks, and coastal rocks.*

The small, fleshy leaves and succulent stems of Sea Milkwort help the plant to conserve fresh water, allowing it to adapt well to the salty conditions of its coastal habitat. This plant tends to grow low, hugging the ground. The solitary flowers are tucked into the leaf bases towards the end of the shoots, forming a prostrate flowering spike.

PERENNIAL

solitary flower at leaf base

succulent leaf

leaf to 1.2cm long

petal-like pink sepals

SIZE *Height to 20cm; flowers 3–6mm wide.*
FLOWERING TIME *May–September.*
LEAVES *Lower leaves alternate, upper leaves opposite; elliptical.*
FRUIT *Small, round capsule at the leaf base.*
DISTRIBUTION *Western Mediterranean, Atlantic, Baltic.*
SIMILAR SPECIES *Bog Pimpernel (p.37), which does not have fleshy leaves.*

Greater Sea-spurrey

FORMS *colonies on upper salt marshes and damp, sandy habitats; occasionally found on coastal rocks.*

Spergularia media (Caryophyllaceae)

A creeping plant, Greater Sea-spurrey is inconspicuous, until its large, pale pink flowers open and reveal the full extent of its spread. It is distinguished from several similar looking species by the pale colour of its petals, and especially, by the green sepals, which are distinctly shorter than the petals.

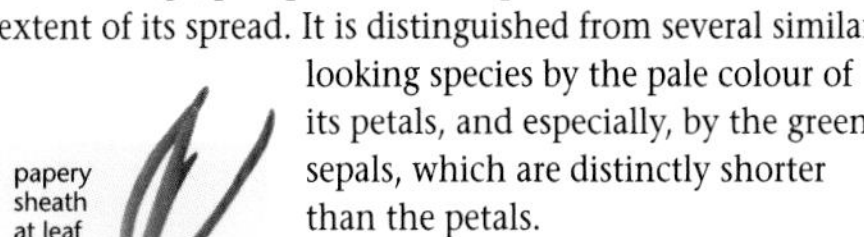

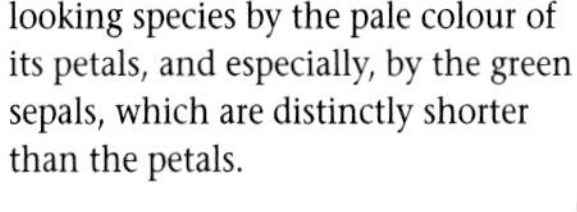

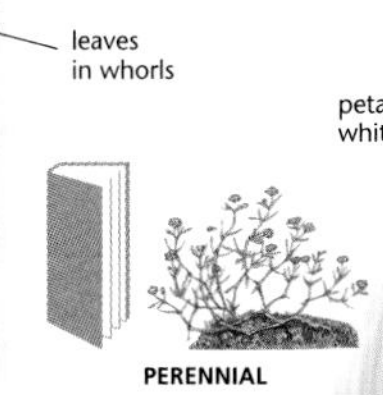

SIZE *Height to 20cm; flowers 0.7–1.2cm wide.*
FLOWERING TIME *May–September.*
LEAVES *Linear, whorled, slightly fleshy.*
FRUIT *Three-valved capsule, on drooping stalk.*
DISTRIBUTION *Mediterranean, Atlantic, southern Baltic.*
SIMILAR SPECIES S. marina, *which has smaller, deep pink flowers;* S. rubra, *which is not fleshy;* S. rupicola, *which has sticky stems.*

Common Stork's-bill

FOUND *in dry, sandy, disturbed, and trampled coastal habitats, including dunes and shingle.*

Erodium cicutarium (Geraniaceae)

Its long-beaked fruit has earned this plant its name – Common Stork's-bill. When the fruit's seeds are shed, the style remains attached to the seed and curls into a spiral as it dries. Then, as rehydration occurs after rain, the style uncoils, drilling the seed into the ground where it can germinate. Biennial plants overwinter as a neat, dark green rosette of leaves.

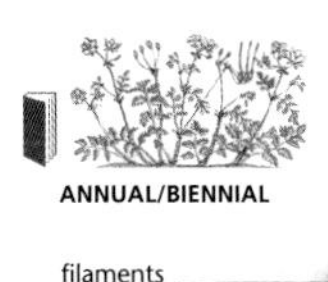

SIZE *Height to 40cm; flowers 1–1.8cm wide.*
FLOWERING TIME *June–September.*
LEAVES *Basal, alternate, pinnate.*
FRUIT *Five mericarps, joined into a beak.*
DISTRIBUTION *Throughout the region.*
SIMILAR SPECIES E. maritimum, *which lacks petals and has rounded, shallow-toothed leaves;* E. moschatum, *which is larger and has more coarsely divided leaves.*

Sand Campion

Silene colorata (Caryophyllaceae)

OCCUPIES *upper sandy beaches, fore-dunes, and dry rocky coastal areas; stony habitats inland.*

One of the most dramatic springtime sights is a sandy beach dominated by the flowers of Sand Campion, especially when set alongside the silvery foliage and yellow flowers of Sea Medick (p.16). A variable plant, both in size and leaf shape, its deeply notched, bright pink petals make it stand out among the many campions of the Mediterranean area.

ANNUAL

SIZE *Height to 25cm; flowers 1.2–1.8cm wide.*
FLOWERING TIME *February–May.*
LEAVES *Basal and opposite, oval to spoon-shaped, lower leaves stalked.*
FRUIT *Toothed capsule.*
DISTRIBUTION *Mediterranean.*
SIMILAR SPECIES S. littorea *and* S. pendula, *which are sticky and also have shallowly notched petals.*

Winged Sea-lavender

Limonium sinuatum (Plumbaginaceae)

OCCURS *in rocky and sandy areas near the sea; may dominate the edges of dry salt pans. Occasionally inland.*

A favourite of dried flower arrangers, who refer to it as Statice, the upward-facing flowers of Winged Sea-lavender retain their colour well even when dried. The predominant mauve colour comes from the persistent sepals, while the petals are white or cream. The wiry stems are winged, bearing three or four raised lines of tissue running down to the basal rosette of lobed, roughly hairy leaves.

PERENNIAL

mauve sepals
flowers in flat clusters

SIZE *Height to 40cm; flowers 1.2–1.5cm long.*
FLOWERING TIME *April–September.*
LEAVES *Basal rosette, pinnately lobed or toothed, roughly hairy.*
FRUIT *Single-seeded capsule.*
DISTRIBUTION *Mediterranean; Portugal*
SIMILAR SPECIES L. thouinii *(western Mediterranean), which has unwinged stems, and yellow petals within pale blue sepals.*

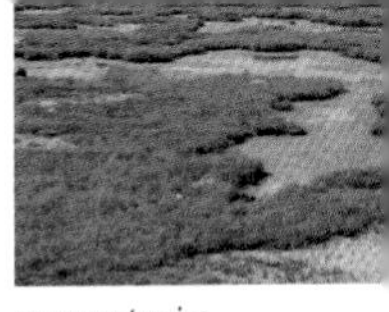

FORMS *extensive carpets on silty, mid-level salt marshes, creating a blanket of colour in mid-summer.*

Common Sea-lavender

Limonium vulgare (Plumbaginaceae)

A variable species in size, flower colour, and habit, the Common Sea-lavender – in all its forms – can transform a salt marsh turf into a carpet of lilac or purple at the height of summer. The small, five-petalled flowers appear in two close-set rows on the upper side of the flowering spike, with each flower set among a series of papery bracts.

SIZE *Height to 40cm; flowers 6–8mm wide.*
FLOWERING TIME *July–October.*
LEAVES *Basal, oblong to elliptical, fleshy.*
FRUIT *Small capsule within a papery calyx.*
DISTRIBUTION *Mediterranean; Atlantic, north to southern Sweden.*
SIMILAR SPECIES *Rock Sea-lavender (below);* L. humile, *which has bluer flowers, and basal branches.*

Rock Sea-lavender

Limonium binervosum agg. (Plumbaginaceae)

FAVOURS *hard and soft coastal rocks and cliffs; some forms grow in areas between sand or shingle and salt marsh.*

Rock Sea-lavender is a confusing complex of numerous closely related species and subspecies. The differences between the forms are relatively slight. The easiest way to identify them is to go by the locality – many forms are unique to, and often the only form found on particular stretches of coastline. All forms, however, have a similar basic appearance.

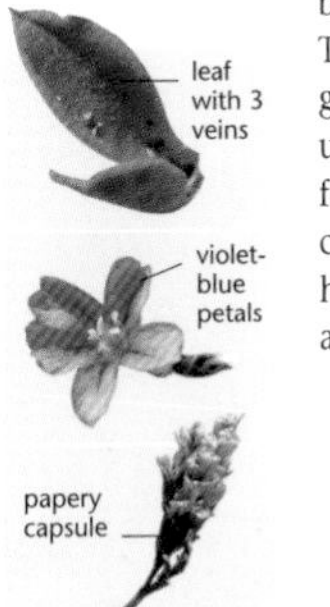

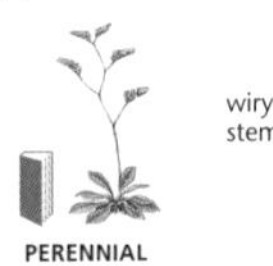

SIZE *Height to 40cm; flowers to 8mm wide.*
FLOWERING TIME *July–September.*
LEAVES *Basal, spoon-shaped to oblong.*
FRUIT *Capsule, with a reddish brown seed.*
DISTRIBUTION *Western Mediterranean; Atlantic, north to Britain.*
SIMILAR SPECIES *Common Sea-lavender (above) has leaves with one vein, and flower spikes that branch at the top of the stem.*

Sea Aster

Aster tripolium (Asteraceae)

OCCUPIES *salt marshes, brackish areas above the tidal influence, also on rocks, cliffs, and inland saline habitats.*

Thriving in all salt marsh zones from seaward, submergent marshes to the extreme high-water mark and above, Sea Aster occurs in two distinct forms. One form, which has many purple ray florets, is characteristic of upper shores. The other form is rayless, with only yellow disc florets, and is usually found closer to the sea – it has air spaces in its tissues to avoid waterlogging. The fleshy, rounded leaves of Sea Aster clasp the stem, which is often reddish in colour. Despite having a pappus for wind dispersal, many seeds are held on the plant until winter, when they become an important food source for finches and other seed-eating birds.

SIZE *Height to 80cm; flowerheads 1–2cm wide.*
FLOWERING TIME *July–October.*
LEAVES *Alternate, linear to lance-shaped.*
FRUIT *Achene with a hairy pappus.*
DISTRIBUTION *Throughout the region.*
SIMILAR SPECIES *Golden Samphire (p.16), which resembles the rayless form;* A. linosyris *(found on southern, limestone cliff slopes), which is also rayless, but has numerous, very narrow, non-succulent leaves; and* A. novi-belgii*, which grows inland.*

NOTE

Sea Aster is a magnet for butterflies. On the coast, it is an invaluable source of nectar for migrating species such as Painted Lady and Red Admiral.

Viper's-bugloss

Echium vulgare (Boraginaceae)

A stunning plant with pink buds, blue flowers, pink stamens, and blue pollen, Viper's-bugloss is instantly recognizable. The flower clusters on the short branches unfurl progressively through the flowering season, so individual plants remain in bloom for several months. The whole plant is bristly hairy, the stem bristles being especially stout, acting as a good deterrant to grazing animals.

GROWS *on shingle banks and sand dunes, often only sparsely vegetated, and in cliff-top grassland.*

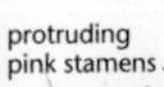

SIZE *Height to 1m; flowers 1.5–2cm long.*
FLOWERING TIME *April–September.*
LEAVES *Basal, alternate, and unstalked.*
FRUIT *Four nutlets at the base of the calyx.*
DISTRIBUTION *Throughout, except parts of the northwest and a few Mediterranean islands.*
SIMILAR SPECIES *Other* Echiums *(largely Mediterranean) such as* E. angustifolium, E. arenarium, *and* E. plantagineum.

Oyster-plant

Mertensia maritima (Boraginaceae)

Hairless, fleshy, and greyish in colour, Oyster-plant is a distinctive and characteristic perennial of northern drift lines. It is generally prostrate, with stems trailing for 80cm or more, and the leaves are arranged in two rows up the stem. The bell-shaped flowers are pink when they open, but turn blue later. Its name is derived from the taste of its leaves, which is reputedly like oysters.

INHABITS *northern coastal sands and shingles. Declining, especially in southern populations.*

PERENNIAL

SIZE *Height to 25cm; flowers 5–7mm wide.*
FLOWERING TIME *June–August.*
LEAVES *Ovate, fleshy, in two rows; lower leaves stalked, upper leaves unstalked.*
FRUIT *Fleshy, flattened nutlets, the outer coat becoming inflated and papery.*
DISTRIBUTION *Atlantic, from northern Britain to the Arctic Circle.*
SIMILAR SPECIES *None.*

Sea Pea

Lathyrus japonicus (Fabaceae)

INHABITS *the less vegetated, seaward areas of shingle banks, and occasionally fore-dunes; very vulnerable to trampling.*

Widely distributed throughout the northern hemisphere, Sea Pea forms loose mats of fleshy, pinnate, blue-green leaves draped over shingle banks. The prostrate, unwinged stems can grow up to 1 metre or more in length. The flowers, borne in loose clusters of up to 12 flowers in a short spike, are purple when they open, fading to blue as they mature. These are followed by hairless, swollen seed pods, which contain the developing seeds. As with cultivated peas, both the young shoots and the seeds are favourite foods of pigeons and doves; the leaflets often have regular peck marks along their margins.

PERENNIAL

NOTE

The seeds of Sea Pea are similar to garden peas, and are edible. A diet of its seeds and Sea Kale (p.24) is reported to have kept entire communities on the Suffolk coast alive during the famines of the Middle Ages.

SIZE *Height to 20cm; spread to 2m; flowers 1.5–2.5cm long.*
FLOWERING TIME *June–August.*
LEAVES *Pinnate; 3–4 pairs of ovate leaflets and unbranched or branched tendrils; broad, pointed stipules.*
FRUIT *Hairless pod, 3–5cm long, containing 4–8 pea-like seeds.*
DISTRIBUTION *Atlantic, north from northern France; Baltic.*
SIMILAR SPECIES Vicia bithynica, *often found in coastal grassland, which has solitary or paired, bicoloured flowers, 1–3 pairs of leaflets, and large, toothed stipules.*

Sea Holly

Eryngium maritimum (Apiaceae)

Living on freely draining sand deposits, Sea Holly has adapted itself well to survive drought-like conditions. It has a waxy, bluish coating on its leaves to reduce water loss, as well as a deep tap-root to store water. It is also very spiny, which deters animals on the lookout for moisture-laden plants, and often develops strong purple pigmentation in the summer as a sunscreen. The spiky, rounded flowerhead is thistle-like, and surrounded by a ruff of spiny, leaf-like bracts. The small flowers are deep blue when freshly open.

FAVOURS *sand dunes, and occasionally shingle banks, often appearing in large colonies; visible from a considerable distance.*

SIZE *Height to 60cm; flowerheads 1.5–3cm wide.*
FLOWERING TIME *June–September.*
LEAVES *Basal and alternate, rounded and lobed, with spiny teeth on the tips.*
FRUIT *Mericarp with overlapping scales.*
DISTRIBUTION *Mediterranean, Atlantic, southern Baltic.*
SIMILAR SPECIES: *Sea Fennel (p.29); other* Eryngium *species can be coastal, especially* E. campestre, *which lacks the waxy, bluish leaf coating and has smaller, greenish flowerheads.*

NOTE

Sea Holly's root was once popular as a candied sweetmeat; this plant has been immortalized in Colchester, Essex, as town-centre fountain.

Hare's-tail Grass

Lagurus ovatus (Poaceae)

OCCURS *on sand dunes and other southern sandy and stony habitats, especially coastal, but also inland.*

Easily recognized by its silvery, egg-shaped flower spikes, Hare's-tail Grass is just as readily identified by touch – the spikes are encased in a soft, silky "fur" of awns, filament-like projections from the individual tiny flowers. A favourite of dried flower arrangers, Hare's-tail Grass is often dyed for flower arrangements.

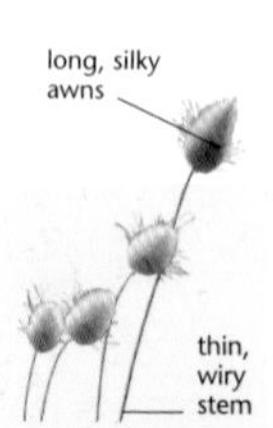

SIZE *Height to 50cm; flowerheads to 2cm long.*
FLOWERING TIME *March–July.*
LEAVES *Soft and downy, to 1cm wide.*
FRUIT *Seeds, in the upper part of the spike.*
DISTRIBUTION *Mediterranean; Atlantic, north to Channel Isles, casual further north.*
SIMILAR SPECIES Cynosurus echinatus, *which has rigid, one-sided spikes;* Polypogon monspeliensis, *which has longer, soft spikes.*

Lyme-grass

Leymus arenarius (Poaceae)

OCCUPIES *seaward sand dunes and upper beaches, and is a primary dune builder.*

A blue-green, broad-leaved plant, Lyme-grass is a characteristic early dune builder. Its flower spikes consist of many overlapping, very flattened spikelets. Two grasses that often grow along with it also have flattened, but shorter spikelets, which are not in opposite pairs: *Elytrigia juncea,* in which the spikelets barely overlap each other, and *E. atherica,* in which they are very close together.

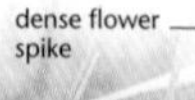

SIZE *Height to 1.5m; spikes to 35cm long; spikelets 1.5–2cm long.*
FLOWERING TIME *July–August.*
LEAVES *Rigid, 0.8–1.5cm wide, short hairs on the veins.*
FRUIT *Starchy grain.*
DISTRIBUTION *Atlantic, north from northwest Spain; Baltic.*
SIMILAR SPECIES *Elytrigia juncea, E. atherica.*

Marram-grass

Ammophila arenaria (Poaceae)

The most important dune-building grass worldwide, Marram-grass is capable of almost unlimited horizontal and vertical growth through deposited sand. The meshwork of roots, shoots, and rhizomes stabilizes the sand – this allows many other plants, including a number of other smaller grasses, such as *Festuca arenaria* and *Vulpia fasciculata*, to colonize the habitat. Protected from erosion by the growth of Marram-grass, sand dunes can build up to a height of more than 100m in areas with a sufficient supply of blown sand, usually where the prevailing winds are onshore.

COLONIZES *low sand dunes, stabilizing the sand and allowing the dunes to grow in height; often planted to control erosion.*

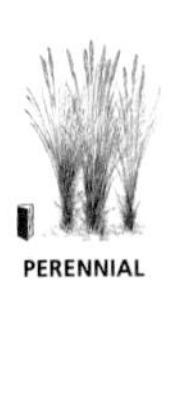

PERENNIAL

clump-forming habit

sharply pointed leaves

SIZE *Height to 1.2m; flower spike to 25cm long; spikelets 1.2–1.4cm long.*
FLOWERING TIME *May–August.*
LEAVES *Tightly inrolled; ribbed above, smooth and shiny below, to 6mm wide.*
FRUIT *Starchy grain.*
DISTRIBUTION *Mediterranean, Atlantic, southern Baltic.*
SIMILAR SPECIES *The hybrid of this plant and* Calamagrostis epigejos, Ammocalamagrostis baltica, *which is more robust, with flatter leaves and often reddish flower spikes.*

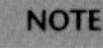

NOTE

Extremely drought tolerant, this plant has waxy-coated leaves with special hinge-cells that wilt before the rest of the plant does. The leaves then roll up tightly to minimize the loss of water to the atmosphere.

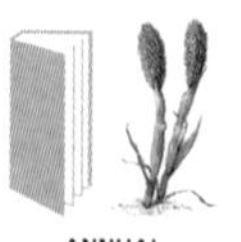

Sand Cat's-tail

Phleum arenarium (Poaceae)

A short grass of sand dunes, Sand Cat's-tail is one of a characteristic group of winter annuals. This group includes Sea Mouse-ear (p.25), the tiny, bright blue-flowered *Myosotis ramosissima*, and the grasses *Parapholis incurva* and *Catapodium marinum*. These plants avoid summer drought by germinating in autumn, flowering in spring, and drying up before the onset of hot weather, surviving as drought-tolerant seeds.

FOUND *on coastal sand dunes and shingle; in sandy areas and locally on mountains inland.*

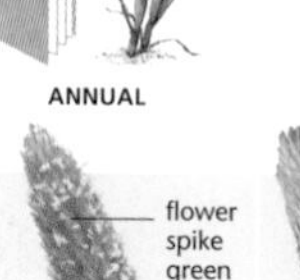

ANNUAL

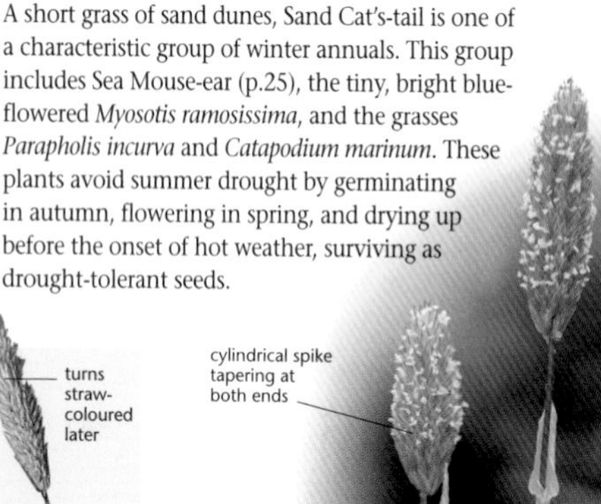

SIZE *Height to 20cm; flower spike to 5cm long.*
FLOWERING TIME *April–June.*
LEAVES *Short and flat, rough on the veins; upper leaf sheaths inflated.*
FRUIT *Small, starchy grain.*
DISTRIBUTION *Western Mediterranean; Atlantic, as far north as Sweden.*
SIMILAR SPECIES *Replaced by* P. graecum *in the eastern Mediterranean.*

Early Sand-grass

Mibora minima (Poaceae)

One of the smallest European grasses, Early Sand-grass is often only 2cm tall. This, together with its early flowering period, makes it inconspicuous and easily overlooked. The spikelets are almost unstalked and arranged in two rows on one side of the slender, hair-like, flower axis. With leaves which are mostly basal, it forms small tufts on bare ground.

OCCURS *in sand dunes and bare sandy ground by the sea, and short, windswept cliff-tops; also sandy areas inland.*

ANNUAL

SIZE *Height to 15cm; spikes to 2cm long; spikelets to 3mm long.*
FLOWERING TIME *February–May.*
LEAVES *Mostly basal, to 0.5mm wide.*
FRUIT *Single starchy grain in each spikelet.*
DISTRIBUTION *Mediterranean; Atlantic, north to Scotland.*
SIMILAR SPECIES Cynodon dactylon, *which has smaller spikelets in 4–6 spreading spikes.*

Grey Hair-grass

Corynephorus canescens (Poaceae)

The strongly tufted Grey Hair-grass imparts a purplish glow to its dune habitats as a result of the pinkish to purple coloration of its spikes and leaf sheaths. The plant is only found in acidic areas, which lack lime. It is, therefore, absent from dunes formed from sand containing the remains of seashells. The bristle-like awn of the flower spike is distinctively thickened at the tip.

PREFERS *more stable parts of acidic sand dunes and shingle; heaths and similar habitats inland.*

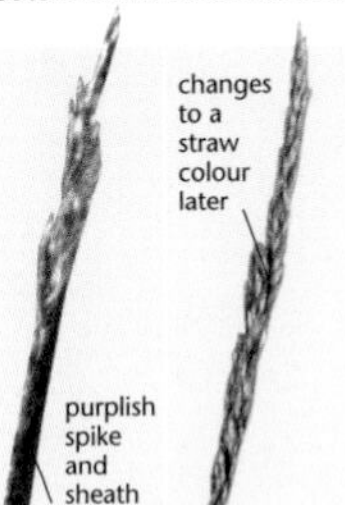

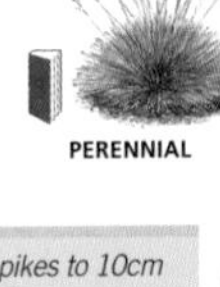

PERENNIAL

tufted growth form

SIZE *Height to 60cm; flower spikes to 10cm long; spikelets 3–4mm long.*
FLOWERING TIME *June–July.*
LEAVES *Hair-like, rigid, with purplish sheaths.*
FRUIT *Starchy grains, two in each spikelet.*
DISTRIBUTION *Western Mediterranean, Atlantic, southern Baltic.*
SIMILAR SPECIES Deschampsia flexuosa *and* Agrostis curtisii, *which are not purplish.*

Common Saltmarsh-grass

Puccinellia maritima (Poaceae)

The tufts of Common Saltmarsh-grass usually merge into a continuous turf as a result of the plant's spreading growth. Although it is usually the dominant species, the turf may contain a number of unrelated plants with superficially similar leaves, including Thrift (p.35), *Plantago maritima,* and *Triglochin maritima;* all three have fleshy leaves.

OCCUPIES *mid- to upper-level salt marshes; saline areas, including road verges, inland.*

PERENNIAL

erect flower branches

narrow leaf, flat or folded

spikelets often purple-tinged

SIZE *Height to 50cm; flower spike to 25cm long; spikelets 0.7–1.2cm long.*
FLOWERING TIME *June–July.*
LEAVES *Smooth, narrow; usually blunt.*
FRUIT *Starchy grain.*
DISTRIBUTION *Atlantic, southern Baltic.*
SIMILAR SPECIES P. distans, *which has reflexed flower branches;* P. fasciculata *and* P. rupestris *have one-sided flower spikes.*

Common Cord-grass

Spartina anglica (Poaceae)

DOMINATES *salt marshes, especially at the lower levels, and encroaches onto upper level mudflats.*

Derived from hybridization a century ago, Common Cord-grass or Rice-grass exhibits hybrid vigour and the ecological tolerance of both its parents. These, together with enhanced fertility due to chromosome doubling, have led to its invasive, spreading growth at the expense of other salt marsh plants. It is often planted to control coastal erosion.

PERENNIAL

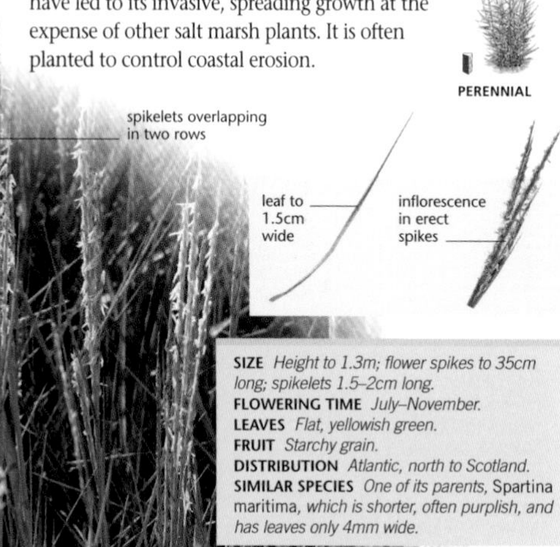

SIZE *Height to 1.3m; flower spikes to 35cm long; spikelets 1.5–2cm long.*
FLOWERING TIME *July–November.*
LEAVES *Flat, yellowish green.*
FRUIT *Starchy grain.*
DISTRIBUTION *Atlantic, north to Scotland.*
SIMILAR SPECIES *One of its parents,* Spartina maritima, *which is shorter, often purplish, and has leaves only 4mm wide.*

Sharp Rush

Juncus acutus (Juncaceae)

FAVOURS *dune slacks, upper salt marshes, and grazing marshes; often found around salt pans.*

With leaves and bracts tipped with exceedingly sharp spines, Sharp Rush lives up to its name. The plant forms substantial clumps, more than a metre tall and wide, which are visible even from a considerable distance. The flowers appear to be borne laterally on the stems, as they are usually overtopped by a bract arising from the base of the flowerhead.

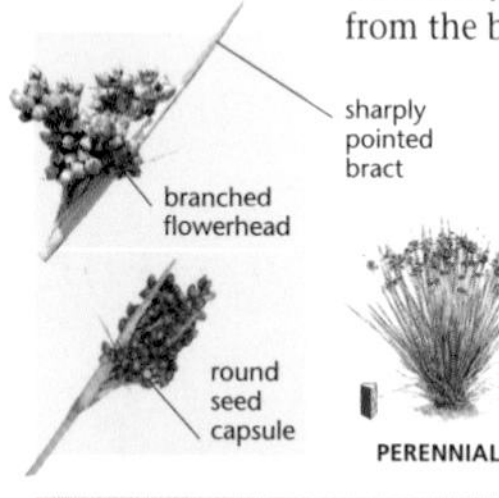

PERENNIAL

SIZE *Height to 1.8m; flowerheads to 25cm long; flower parts to 4mm long.*
FLOWERING TIME *June–July.*
LEAVES *Cylindrical, containing pith.*
FRUIT *Large, pointed, reddish brown capsule.*
DISTRIBUTION *Mediterranean; Atlantic, as far north as Ireland and Wales.*
SIMILAR SPECIES J. balticus, *which is tufted;* J. gerardii, *which is patch-forming.*

Sea Club-rush

Bolboschoenus maritimus (Cyperaceae)

As with many of the sedge family, the stems of the Sea Club-rush are sharply triangular in section, the edges being rough in the upper part of the stem. In flower, its clusters of egg-shaped spikelets, with leaf-like bracts much longer than the flowerhead, are distinctive. Usually found in brackish water, its seeds can be transported to inland ponds and reservoirs by wildfowl.

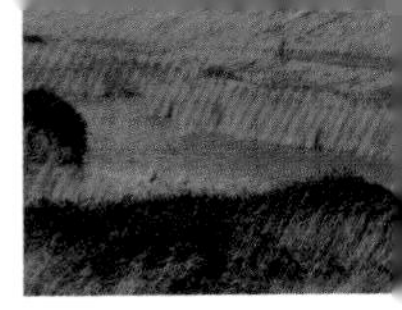

ROOTS *in shallow water, usually brackish; in tidal rivers, ditches, and ponds.*

SIZE *Height to 1.2m; flowerheads to 5cm long; spikelets 1–2cm long.*
FLOWERING TIME *July–August.*
LEAVES *Strongly keeled and rough on the margins and midrib; to 2cm wide.*
FRUIT *Blackish brown nutlet.*
DISTRIBUTION *Throughout the region.*
SIMILAR SPECIES Cyperus capitatus, *which has 4–5 long bracts around a flowerhead.*

Round-headed Club-rush

Scirpoides holoschoenus (Cyperaceae)

A tufted perennial of damp habitats, Round-headed Club-rush has tightly packed, rounded flowerheads, some unstalked and others on stalks to 8cm long. The whole inflorescence appears to be borne laterally on the stem, which in fact is the bract forming a stem-like extension above the flowerheads.

INHABITS *dune slacks and other damp, sandy areas, and grazing marsh ditches; locally in damp places inland.*

PERENNIAL

stem-like bract

bract above inflorescence

variably stalked flower heads

SIZE *Height to 1.5m; flowerheads to 1.5cm wide.*
FLOWERING TIME *June–September.*
LEAVES *Basal, strap-like, often absent.*
FRUIT *Triangular, pale brown nutlet.*
DISTRIBUTION *Mediterranean; Atlantic, as far north as southern Britain.*
SIMILAR SPECIES Schoenoplectus tabernaemontani, *which has oval flowerheads.*

Common Reed

Phragmites australis (Poaceae)

THRIVES *in fresh to brackish water, to one-third the salinity of sea water, in upper salt marshes, lagoons, and grazing marsh ditches, often in large stands.*

The extensively creeping rhizomes of the Common Reed results in dense reedbeds, which are an important habitat for a wide range of other plants and animals. The nodding, purplish flowers contain silky hairs that are especially conspicuous when fruiting; caught by low autumn sunlight, they create a beautiful golden glow over vast areas of wetland.

SIZE *Height to 3.5m; flower spike to 30cm long; spikelets 1–1.5cm.*
FLOWERING TIME *August–October.*
LEAVES *Flat, grey-green; to 5cm wide.*
FRUIT *Starchy grain.*
DISTRIBUTION *Mediterranean; Atlantic, north to southern Norway; Baltic.*
SIMILAR SPECIES Arundo donax *is taller (to 6m or more), and is found by the Mediterranean.*

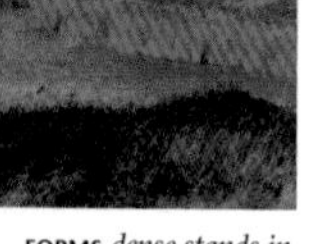

Lesser Reedmace

Typha angustifolia (Typhaceae)

FORMS *dense stands in brackish waters near the coast, in grazing marshes and lagoons; also occurs inland.*

Often incorrectly called a Bulrush, the Lesser Reedmace often replaces the widespread Greater Reedmace (*T. Latifolia*) in coastal and brackish localities, although both are also found together. The flower spikes are usually clearly separated – male above and female below. The seeds are held in the spike until winter, when they form an important food resource for reedbed birds.

SIZE *Height to 2m; female spike to 15cm long and 2.5cm wide; male spike narrower.*
FLOWERING TIME *June–August.*
LEAVES *Flat, to 6mm wide.*
FRUIT *Small, one-seeded, dry capsule, producing a cottony down when ripe.*
DISTRIBUTION *Throughout the region.*
SIMILAR SPECIES T. latifolia, *which has leaves more than 8mm wide.*

Sand Sedge

Carex arenaria (Cyperaceae)

The rhizome of Sand Sedge is tipped with a hardened sheath to protect the growing point as it advances through abrasive sand. Shoots are produced along the rhizome, growing above the ground in straight lines. By varying the distance between shoots, and the branching angle and frequency of the rhizomes, Sand Sedge is able to congregate its shoots in optimal habitats.

SPREADS *through sandy coastal habitats, especially sand dunes, which it colonizes rapidly.*

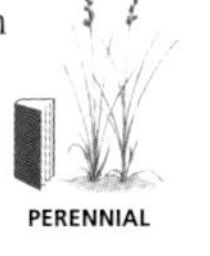

pale brown flowerhead, with bracts

flat leaves, sometimes inrolled

SIZE *Height to 40cm; flower spike to 8cm long.*
FLOWERING TIME *June–July.*
LEAVES *Flat or inrolled, to 4mm wide.*
FRUIT *Yellow-green beaked nutlet.*
DISTRIBUTION *Western Mediterranean, Atlantic, southern Baltic.*
SIMILAR SPECIES Carex divisa, *which has longer bracts;* Carex maritima, *which has almost globular flowerheads.*

Long-bracted Sedge

Carex extensa (Cyperaceae)

A rigid, tufted perennial, Long-bracted Sedge has male and female flowers in separate spikes. The single male spike is surrounded by several broader female spikes, usually clustered at the top of the stem, although sometimes the lowest female spike is stalked and placed well down the stem. The lower spikes have long bracts, which overtop the whole inflorescence by a considerable margin.

FAVOURS *upper, drier salt marshes and bare areas near the sea; also on coastal cliffs and rocks.*

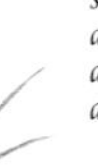

SIZE *Height to 40cm; male spike to 2.5cm long; female spikes to 2cm long.*
FLOWERING TIME *June–July.*
LEAVES *Stiff, grooved, to 3mm wide.*
FRUIT *Green or brown, long-beaked nutlet.*
DISTRIBUTION *Mediterranean; Atlantic, north to Scotland; southern Baltic.*
SIMILAR SPECIES Carex distans *has female spikes that do not, or only slightly, overlap.*

Neptune Weed

Posidonia oceanica (Posidoniaceae)

FORMS *dense meadows in shallow water with a sandy bottom, creating an important habitat for fish and marine invertebrates.*

One of the few higher plants to thrive in salt water, Neptune Weed forms extensive meadows in clear water through which sunlight can penetrate. It has grass-like leaves, and small clusters of insignificant flowers that lack petals. It is most familiar as a deposit of dead leaves and fibrous balls (see p.209) stranded upon nearby beaches.

PERENNIAL

grass-like leaves in shallow water

leaf to 1cm wide

stem-base clothed in fibrous leaf remains

SIZE *Height in water to 80cm.*
FLOWERING TIME *October–March.*
LEAVES *Strap-like, blunt or notched at the tip.*
FRUIT *Fleshy berry, also called Sea Olive.*
DISTRIBUTION *Mediterranean.*
SIMILAR SPECIES *Common Eel-grass (below);* Cymodocea nodosa, *which has leaves to 4mm wide, with fewer veins and reddish leaf sheaths.*

Common Eel-grass

Zostera marina (Zosteraceae)

GROWS *on sheltered intertidal mud- and sandflats, extending below the low tide mark in clear water.*

Intertidal and sub-tidal meadows of Common Eel-grass are an important component of the marine ecosystem, affording shelter to a wide range of animals. The leaves are a significant food resource for grazing wildfowl such as Brent Geese (p.186), while the creeping rhizomes help to bind and stabilize the sediment in which the plant grows.

dark green leaf, to 1cm wide

flaccid, grass-like leaf

PERENNIAL

SIZE *Leaves to 50cm long; inflorescence 9–12cm long.*
FLOWERING TIME *June–September.*
LEAVES *Grass-like, a point extending from the rounded leaf tip.*
FRUIT *Ribbed, oval seed, to 3mm long.*
DISTRIBUTION *Throughout the region.*
SIMILAR SPECIES Z. noltii *has leaves about 1mm wide, and has a more estuarine habitat.*

Fennel Pondweed

Potamogeton pectinatus (Potamogetonaceae)

One of a group of unrelated water plants with very narrow leaves, Fennel Pondweed is easiest to distinguish when in flower. Its small, greenish flowers form a whorled spike, the whorls usually separate from one another, especially when in fruit. The flowers may be produced underwater or above the water surface. The translucent leaves have rounded to pointed tips.

FREQUENTS *standing or slow-moving water in grazing marshes; also widespread inland in lowland areas.*

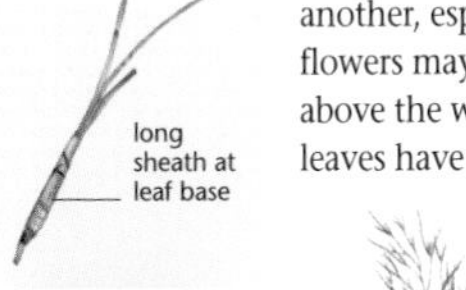

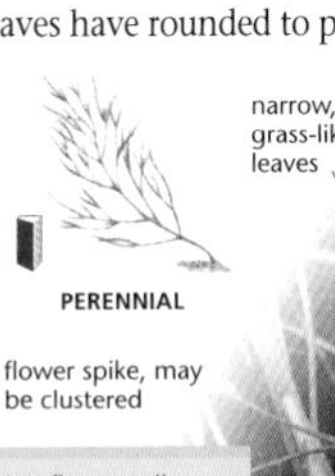

SIZE *To 2m long underwater; flower spike 2–5cm long.*
FLOWERING TIME *May–September.*
LEAVES *Grass-like, with 2 hollow tubes.*
FRUIT *Asymmetrical, beaked achene.*
DISTRIBUTION *Throughout the region.*
SIMILAR SPECIES *Beaked Tasselweed (below);* Zannichellia palustris, *which has solid leaves and flowers enclosed by the leaf sheath.*

Beaked Tasselweed

Ruppia maritima (Ruppiaceae)

As with many aquatic plants, the flowers of Beaked Tasselweed are relatively simple. They lack petals and consist of just two stamens and four carpels. As the fruit forms, each develops a long stalk, forming an umbel-like structure at the end that usually reaches the surface of water. This unique flower structure is an important identification feature, distinguishing it from several other similar water plants.

FAVOURS *brackish water bodies, especially coastal ditches and lagoons.*

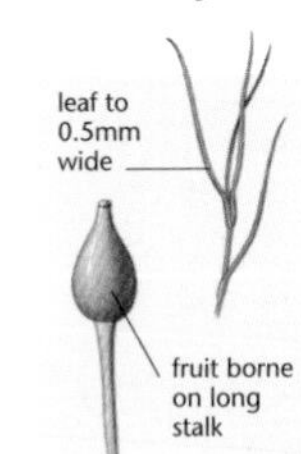

SIZE *Length to 40cm underwater; tiny flowers, in umbels on stalks to 6cm long when fruiting.*
FLOWERING TIME *July–September.*
LEAVES *Narrow, thread-like, bright green.*
FRUIT *Pear-shaped, beaked, fleshy drupe.*
DISTRIBUTION *Throughout the region.*
SIMILAR SPECIES R. cirrhosa, *which has slightly broader leaves, and the flower stalk is coiled when in fruit.*

Pirri-pirri Bur

Acaena novae-zelandiae (Rosaceae)

COLONIZES *bare lowland habitats, especially sand dunes, in which it can become locally dominant.*

A native of New Zealand and Australia, Pirri-pirri Bur is one of a number of *Acaena* species that has colonized Europe, largely as garden escapes or as seed in imported wool. The fruiting heads are admirably suited to dispersal on wool or clothes, each flower bearing hooked spines that attach themselves to animals or clothing. Unfortunately, this ensures its rapid spread, to the detriment of native species.

toothed leaflet

reddish fruit in autumn

SIZE *Height to 15cm; flowerheads to 1cm wide (excluding spines).*
FLOWERING TIME *July–August.*
LEAVES *Pinnate; 7–9 pairs of toothed leaflets.*
FRUIT *Globular bur, with long, barbed bristles.*
DISTRIBUTION *Britain and Ireland.*
SIMILAR SPECIES *Several other* Acaena *species, which are less frequent.*

Spiny Cocklebur

Xanthium spinosum (Asteraceae)

THRIVES *in disturbed lowland habitats, close to the sea; also found on beaches and strandlines.*

Originating from South America, Spiny Cocklebur is now a familiar species of the coastal fringe of the Mediterranean basin. The presence of yellowish spines, usually three-forked, at the base of each leaf, helps distinguish it from the related Rough Cocklebur (*X. strumarium*). It also bears globular, greenish male flowerheads to 6mm wide, with the female flowers growing in clusters from the leaf axil.

ANNUAL

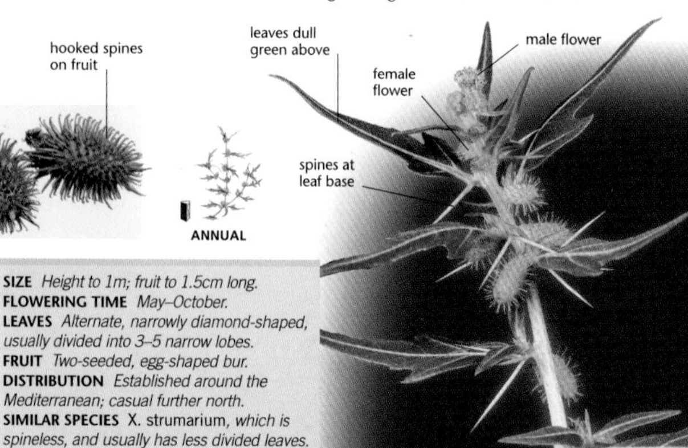

SIZE *Height to 1m; fruit to 1.5cm long.*
FLOWERING TIME *May–October.*
LEAVES *Alternate, narrowly diamond-shaped, usually divided into 3–5 narrow lobes.*
FRUIT *Two-seeded, egg-shaped bur.*
DISTRIBUTION *Established around the Mediterranean; casual further north.*
SIMILAR SPECIES X. strumarium, *which is spineless, and usually has less divided leaves.*

Frosted Orache

Atriplex laciniata (Chenopodiaceae)

A sprawling plant, Frosted Orache is recognizable from a distance by its silvery green colour. Its drift-line habitat is often shared with related annual oraches and goosefoots, though none is as silvery, and most are also found inland. Stinking Goosefoot (*Chenopodium vulvaria*) is similarly mealy and sprawling, but its leaves are less lobed, and it has a distinct stench of rotting fish.

GROWS *on and just above the high-water mark on sandy and shingle shores.*

cluster of tiny flowers lacking petals

ANNUAL

SIZE *Height to 30cm; flower spike to 2cm long.*
FLOWERING TIME *July–September.*
LEAVES *Alternate, diamond-shaped, lobed.*
FRUIT *Achene inside expanded, toothed bracteole.*
DISTRIBUTION *Atlantic, southern Baltic.*
SIMILAR SPECIES A. patula, prostrata, *and* littoralis, *which have diamond-shaped, broadly triangular, and linear leaves, respectively.*

Sea Beet

Beta vulgaris maritima (Chenopodiaceae)

There is little in the appearance of this plant to show that it is the forerunner to the modern beetroot, except that the glossy, fleshy leaves and stems are often red-tinged. It has a very prostrate habit, with long, trailing flowering stems. The tiny greenish flowers are borne in clusters of three on leafy spikes.

SPRAWLS *over shingle beaches, salt marshes, cliffs, old sea walls, and grassy embankments; often close to tide line.*

ANNUAL/PERENNIAL

slender flower spike

untoothed margin

flowers in small clusters

long leaf stalk

reddish stem

SIZE *Height to 1m, flowers 2–4mm wide*
FLOWERING TIME *June–September.*
LEAVES *Alternate, fleshy.*
FRUIT *Corky, swollen segments.*
DISTRIBUTION *Mediterranean; Atlantic, north to Denmark.*
SIMILAR SPECIES *The Mediterranean* B. macrocarpa, *which has flower spikes that are leafy right to the top.*

Prickly Saltwort

Salsola kali (Chenopodiaceae)

OCCURS *on sandy coastal beaches, or in shingle, often close to the tideline.*

This plant is recognizable by a sharp spine on the tip of each leaf, unusual for a plant that is beyond the reach of grazing animals. Prickly Saltwort is succulent, branched, and bluish green. Its tiny yellow flowers, hidden at the base of the fleshy leaves, may have a pinkish tint, but like many of the spinach family, they are rather insignificant.

ANNUAL

SIZE *Height to 80cm, flower 2–3mm wide.*
FLOWERING TIME *July–October.*
LEAVES *Alternate, linear to oval, succulent, spine-tipped.*
FRUIT *Achene covered by flower parts.*
DISTRIBUTION *Mediterranean; Atlantic, north to southern Norway; southern Baltic.*
SIMILAR SPECIES Corispermum leptopterum, *which has flowers in cone-like heads.*

Purple Spurge

Euphorbia peplis (Euphorbiaceae)

INHABITS *the upper parts of sand and shingle beaches, just above the strandline.*

A creeping plant that rarely grows taller than 4cm, Purple Spurge bears all the hallmarks of its genus. Its flowers are arranged in cyathia (one female and several male flowers grouped within a cup-shaped bract, containing nectar glands). The stems exude a milky latex when broken. It is an early colonizer of sand deposits, but is now extinct in Britain and the Channel Islands.

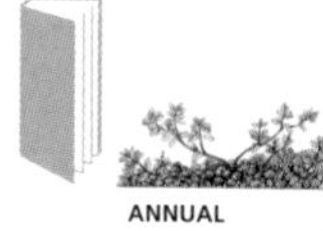

ANNUAL

SIZE *Height to 4cm; spread around 20cm; flower clusters to 3mm wide.*
FLOWERING TIME *July–September.*
LEAVES *Opposite; oblong, with a rounded lobe at one side of the base; to 1cm long.*
FRUIT *Three-lobed, rounded capsule.*
DISTRIBUTION *Mediterranean; Atlantic, as far north as France (extinct in Britain).*
SIMILAR SPECIES *None.*

Sea Spurge

Euphorbia paralias (Euphorbiaceae)

Branched only at the base, Sea Spurge produces a few upright stems. These stems carry dense ranks of oblong or elliptical leaves, all of which fan out horizontally. The whole plant is blue-green in appearance, especially when young. The flowers are produced from the upper leaf axils.

FAVOURS *sand dunes, upper beaches, and other sandy coastal habitats.*

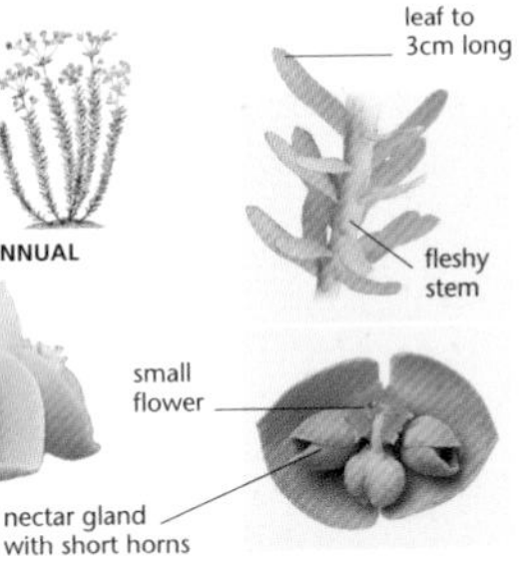

SIZE *Height to 40cm; flower clusters (cyathia) to 1cm wide.*
FLOWERING TIME *May–September*
LEAVES *Oblong or oval.*
FRUIT *Granular capsule; 3–5mm wide*
DISTRIBUTION *Mediterranean; Atlantic, as far north as Northern Ireland.*
SIMILAR SPECIES E. pithyusa, *found in S. France and Italy, which has narrow leaves.*

Portland Spurge

Euphorbia portlandica (Euphorbiaceae)

Very similar to Sea Spurge, with which it often grows, Portland Spurge flowers a little earlier, and its stems and lower leaves often become deep red. Its flower glands have long, curved horns, and its seeds are pitted rather than smooth. A further distinction is that Portland Spurge exploits other coastal habitats, notably cliffs.

FOUND *on sand dunes and upper sandy beaches; on cliff-top edges and coastal rocks.*

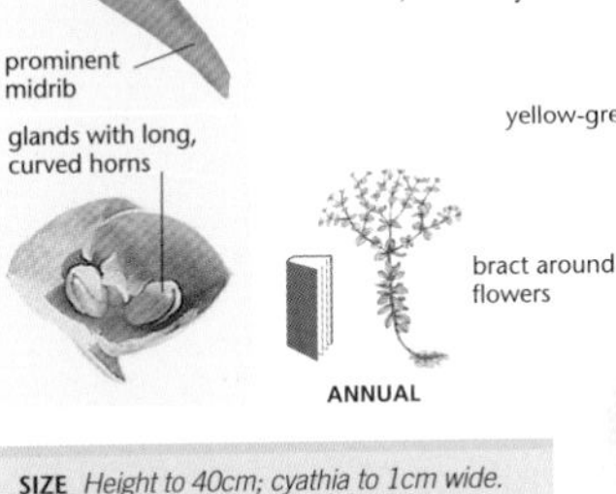

SIZE *Height to 40cm; cyathia to 1cm wide.*
FLOWERING TIME *April–September.*
LEAVES *Oval, waxy, blue-grey.*
FRUIT *Granular capsule, to 3mm wide.*
DISTRIBUTION *Atlantic, as far north as Northern Ireland.*
SIMILAR SPECIES *Sea Spurge (above), which has shorter horns on its flower glands, and smooth seeds.*

Sea-purslane

Atriplex portulacoides (Chenopodiaceae)

THRIVES *in mid- to upper-level salt marshes, especially on channel edges, and salt-splashed cliffs.*

This common salt marsh plant forms very extensive colonies. It has woody stems at the base, which produce a mass of fleshy leaves. The leaves are silvery-grey, glistening with salt crystals released from special glands. Clusters of tiny, yellowish flowers appear in late summer.

SIZE *Height 20–60cm; flowers 2–3mm wide.*
FLOWERING TIME *July–October.*
LEAVES *Opposite, oblong, untoothed, silvery-mealy.*
FRUIT *Single-seeded achene.*
DISTRIBUTION *Mediterranean; Atlantic, north to Denmark.*
SIMILAR SPECIES *Annual Sea-purslane (below).*

Annual Sea-purslane

Atriplex pedunculata (Chenopodiaceae)

PREFERS *upper salt marsh habitats, on the interface zone with sand dunes; also saline habitats inland.*

A declining species throughout its range, Annual Sea-purslane is easily overlooked on account of its late flowering, and the similarity of its leaves to those of its perennial relative, Sea-purslane (above). However, for anyone looking at it in autumn, it is unmistakable, its fruit borne on long stalks, almost like the familiar pods of Shepherd's-purse (*Capsella bursa-pastoris*).

SIZE *Height to 35cm; flowers 1–2mm wide, in spikes to 8cm long.*
FLOWERING TIME *August–October.*
LEAVES *Alternate, oblong, short-stalked.*
FRUIT *Achene, within long-stalked, fused, lobed bracteoles.*
DISTRIBUTION *Atlantic, southern Baltic.*
SIMILAR SPECIES *Diminutive, non-fruiting specimens of Sea-purslane (above).*

Shrubby Sea-blite

Suaeda vera (Chenopodiaceae)

Usually marking the extreme high-water mark on a salt marsh, Shrubby Seablite is a bushy species with woody stems and branches, which contrast with its fleshy leaves. The horizontal branches root readily into the substrate, enabling this plant to form dense colonies. The older branches are frequently colonized by lichens, especially the orange Sunburst Lichen (p.76).

FAVOURS *upper salt marshes; also on shingle, salt-sprayed cliffs, and around saline lagoons and salt pans.*

PERENNIAL

SIZE *Height to 1.2m; flowers 2mm wide.*
FLOWERING TIME *July–October.*
LEAVES *Linear, blue-grey, tinged red at times.*
FRUIT *Oval, blackish, shining achene.*
DISTRIBUTION *Mediterranean, Atlantic.*
SIMILAR SPECIES *Shrubby Glasswort (p.63), which has cylindrical, fleshy stems and branches;* Halocnemum strobilaceum, *which has papery, scale-like leaves.*

Annual Sea-blite

Suaeda maritima (Chenopodiaceae)

An early colonizer of intertidal mud, often with Glasswort (p.62), Annual Sea-blite is a very variable plant – in form (prostrate to erect), colour (dark green to purple), branching habit, and flowering time. Some different forms have been recognized as distinct varieties, especially those prostrate, short-leaved, early-flowering plants (var. *macrocarpa*), which also tend to have larger seeds.

INHABITS *low-level salt marshes, penetrating upper marshes along creeks and pools; also in saline areas inland.*

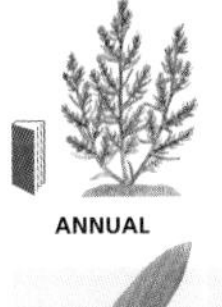

ANNUAL

fleshy, pointed leaf

greenish flower with 2 stigmas

achene to 1.8cm wide

SIZE *Height to 50cm; prostrate forms less than 20cm; flowers 2mm wide.*
FLOWERING TIME *July–October.*
LEAVES *Linear, half-rounded, to 2.5cm long.*
FRUIT *Round, blackish, shining achene.*
DISTRIBUTION *Mediterranean, Atlantic, southern Baltic.*
SIMILAR SPECIES S. splendens *(Mediterranean) has semi-transparent leaves.*

Glasswort

Salicornia europaea agg. (Chenopodiaceae)

OCCURS *across the whole range of salt marsh habitats, from almost bare upper mud-flats to the splash zone above the tidal limit.*

Also known as Marsh Samphire, although unrelated to other samphires (pp.14 and 18), Glasswort is actually a group (or aggregate) of similar, very closely related annual species, found across the full range of salt marsh habitats. All species have a similar succulent form, their cactus-like appearance reflecting the lack of fresh water in their saline habitats. The insignificant flowers are found singly or in groups of three at the base of the segments at the top of the shoot. The seeds of Glasswort are rich in oils, and are a favoured autumn food of Teal (*Anas crecca*) and other coastal ducks.

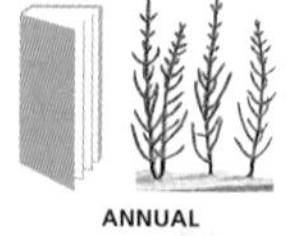

ANNUAL

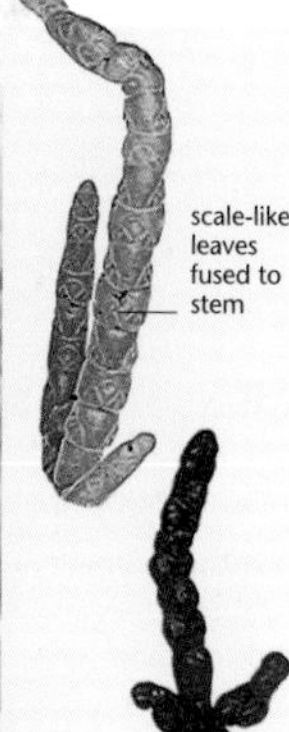

NOTE

Often the easiest way to distinguish the individual species of Glasswort is by their distinctive autumn tints – for example, S. ramosissima *turns deep reddish purple,* S. pusilla *turns orange-pink, and* S. dolichostachya *turns yellow-green during the season.*

SIZE *Height to 30cm; flowers tiny, in spikes to 10cm long, but much shorter in some species.*
FLOWERING TIME *August–September.*
LEAVES *Fused around the stem, forming cylindrical segments; succulent.*
FRUIT *Tiny achene, covered in hooked hairs.*
DISTRIBUTION *Mediterranean, Atlantic, southern Baltic.*
SIMILAR SPECIES Annual Seablite *(p.61), which has fleshy leaves, not fused to the stem;* Perennial Glasswort *(p.63), which is a patch-forming, creeping, woody plant.*

Perennial Glasswort

Sarcocornia perennis (Chenopodiaceae)

Very similar in appearance to its annual relative, Glasswort (p.62), Perennial Glasswort is distinguished mainly by its form – spreading, woody branches, which take root as they grow, forming large, dense patches. It has cylindrical, segmented stems and branches. Its flowers, at the base of the upper stem segments, are tiny and in groups of three.

FORMS *creeping patches on upper salt marshes, and where salt marsh interfaces with shingle.*

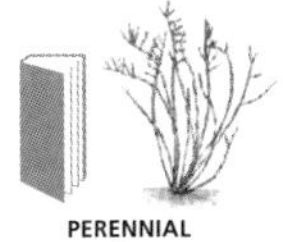

SIZE *Height to 30cm; spread to 1m; flower spikes to 4cm long.*
FLOWERING TIME *August–September.*
LEAVES *Fused around the stem, forming segments; succulent.*
FRUIT *Tiny achene, covered in hooked hairs.*
DISTRIBUTION *Mediterranean; Atlantic, as far north as Britain.*
SIMILAR SPECIES *Glasswort.*

Shrubby Glasswort

Arthrocnemum macrostachyum (Chenopodiaceae)

Growing more erect than Perennial Glasswort (above), Shrubby Glasswort is a characteristic shrub of Mediterranean coastal lowlands, around saline waters. Although it does not spread and take root like Perennial Glasswort, it can form dense thickets, which are often home to typical scrub birds such as the Sardinian Warbler (*Sylvia melanocephala*). Its flowers are in groups of three and the leaves are scale-like.

FAVOURS *the banks of saline lagoons and salt pans, and upper salt marsh habitats.*

SIZE *Height to 1m; flower spikes to 8cm long.*
FLOWERING TIME *May–August.*
LEAVES *Fused around the stem, forming segments; succulent.*
FRUIT *Tiny, black achene, with blunt projections.*
DISTRIBUTION *Mediterranean.*
SIMILAR SPECIES A. fruticosum, *which is more prostrate, and has a later flowering time.*

CREEPS *over extensive, stable shingle deposits, forming ankle-high thickets; the more usual upright form is widespread everywhere.*

Prostrate Blackthorn

Prunus spinosa (Rosaceae)

A common shrub throughout most of Europe, Blackthorn forms dense, impenetrable stands in a range of habitats, including windswept cliff-tops and sand dunes. Such thickets are generally between one and four metres tall. However, on large, stable shingle systems a prostrate variety is characteristic, which is only a few centimetres high. Prostrate plants often have somewhat smaller leaves than those of the shrubby form, though the flowers and fruit are identical.

toothed leaf

several stamens

black fruit, with blue-grey bloom

SIZE *Height to 30cm; flowers to 2cm wide.*
FLOWERING TIME *April–May.*
LEAVES *Oblong, toothed, 2–3cm long, appearing after the flowers open.*
FRUIT *Globose drupe, 1–1.5cm diameter; blue-black, with a waxy bloom.*
DISTRIBUTION *Prostrate form not precisely known, but mainly Atlantic, and especially around the British Isles.*
SIMILAR SPECIES *None with the prostrate habit.* P. cerasifera *is similar to the shrubby form, but its flowers appear earlier, with the leaves, while the ripe fruit are red.*

NOTE

The extreme environment of shingle has produced numerous variants, often prostrate, purple-leaved, and/or succulent, of familiar species, including Cytisus scoparius, Geranium robertianum, *and* Solanum dulcamara.

Japanese Rose

DOMINATES *sand dunes, often forming dense patches; inland on waste ground, and common in gardens.*

Rosa rugosa (Roasaceae)

A native of sand dunes in eastern Asia, the Japanese Rose has escaped from cultivation in northern Europe and invaded similar habitats here. Its dense, spreading habit and salt-tolerant nature mean that it can become quite dominant, to the detriment of native species. Its large, scented flowers are attractive to bees, and its berries are opened in the autumn by seed-eating birds.

PERENNIAL

prickly stems

SIZE *Height to 2m, usually under 1m on sand dunes; flowers 6–8cm wide.*
FLOWERING TIME *June–September.*
LEAVES *Alternate, pinnate; leaflets serrated.*
FRUIT *Large, rounded red hips, to 2.5cm wide.*
DISTRIBUTION *Atlantic, north from northern France, and especially around the Baltic.*
SIMILAR SPECIES *None.*

Burnet Rose

Rosa pimpinellifolia (Rosaceae)

A patch-forming low shrub, spreading by suckers, Burnet Rose is very densely clothed in spines and stiff bristles, hence its former scientific name *Rosa spinosissima*, "the most spiny one of all". Other wild roses, which may be found in similar habitats, usually have fewer, larger, curved thorns, and tend to have a scrambling habit.

FORMS *suckering patches on stable sand dunes and cliff-top heathland.*

PERENNIAL

dead sepals

blackish purple ripe hip

reddish young hip

SIZE *Height to 50cm, though often larger inland; flowers 2–4cm wide.*
FLOWERING TIME *May–July.*
LEAVES *Pinnate, 4–5 pairs of small, round, toothed leaflets.*
FRUIT *Globose hip, to 1.5cm wide.*
DISTRIBUTION *Mediterranean; Atlantic, as far north as southern Norway.*
SIMILAR SPECIES *None.*

Creeping Willow

Salix repens (Salicaceae)

OCCURS *in damp, often winter-wet, dune slacks; occasionally as flushes on soft cliffs.*

A variable dwarf shrub of damp areas, Creeping Willow occurs in several forms: var. *argentea*, which is upright rather than creeping, is especially associated with sand dunes. Its silvery leaves help to reflect the heat of the sun in summer.

SIZE *Height to 1m; catkins to 2.5cm long and 8mm wide.*
FLOWERING TIME *April–June.*
LEAVES *Broad, lance-shaped, with silky hairs.*
FRUIT *Many-seeded capsule, with downy hairs for wind dispersal.*
DISTRIBUTION *Atlantic, Baltic.*
SIMILAR SPECIES *No other willows share the habitat of var.* argentea.

Sea-buckthorn

Hippophae rhamnoides (Elaeagnaceae)

OCCUPIES *large areas of stable sand dunes and soft cliff slopes. Inland on river gravels.*

A dense, thorny shrub, Sea Buckthorn is often planted to stabilize sand dunes. Forming impenetrable thickets, it excludes all other plants, and is consequently considered to be a problem species. However, its berries, which ripen in late summer, are a rich source of food for migrating birds.

SIZE *Height to 3m; flowers clusters to 8mm.*
FLOWERING TIME *March–June.*
LEAVES *Narrow, linear to lance-shaped, silvery when young.*
FRUIT *Round berry, to 8mm wide.*
DISTRIBUTION *Atlantic, north from northern Spain; Baltic. Native distribution obscured by planting to stabilize sand dunes.*
SIMILAR SPECIES *None.*

African Tamarisk

Tamarix africana (Tamaricaceae)

Particularly around the Mediterranean, there are several species of *Tamarix*, which look similar but have slightly different leaves or flowers. However, they all have tiny, blue-green leaves which are slightly pressed close to the branches, giving a feathery appearance. A small tree, African Tamarisk shows its full glory when the flowers emerge in abundant spikes, casting a pinkish glow over the whole canopy. Flowering is predominantly in spring, but sometimes blossoms reappear in the early autumn when the rains return – part of the Mediterranean "second spring".

GROWS *alongside coastal marshes and lagoons, penetrating inland on riverbanks; often planted on upper sandy beaches.*

NOTE

Despite the feathery branches, tamarisks cast useful shade in the heat of the day. Several species are therefore planted on Mediterranean beaches, their trunks whitewashed to help them tolerate the heat.

SIZE *Height to 6m; flowers 4–5mm wide, in drooping spikes to 6cm long.*
FLOWERING TIME *March–July.*
LEAVES *Alternate, pointed, 2–4mm long.*
FRUIT *Capsule with numerous, hairy, wind-dispersed seeds.*
DISTRIBUTION *Western Mediterranean, Atlantic (Portugal). Widely planted further east and north.*
SIMILAR SPECIES *Several other* Tamarix *species, such as* T. gallica *(native north from Brittany, and much planted in coastal sites north to Britain), which has narrower flower spikes.*

Prickly Juniper

Juniperus oxycedrus (Cupressaceae)

FAVOURS *maritime rocks and sandy habitats, often growing twisted and distorted by the effects of salt-laden sea spray.*

A widespread southern European tree, Prickly Juniper occurs in a distinctive maritime variety, the subspecies *J. o. macrocarpa*, which is distinguished by its larger cones and slightly wider needles. It bears tiny flowers in clusters: the male flowers are yellow, while the female ones are green. The berry-like cones take two years to ripen; at first they are green, with a waxy bloom, but later become dark red-brown or purple.

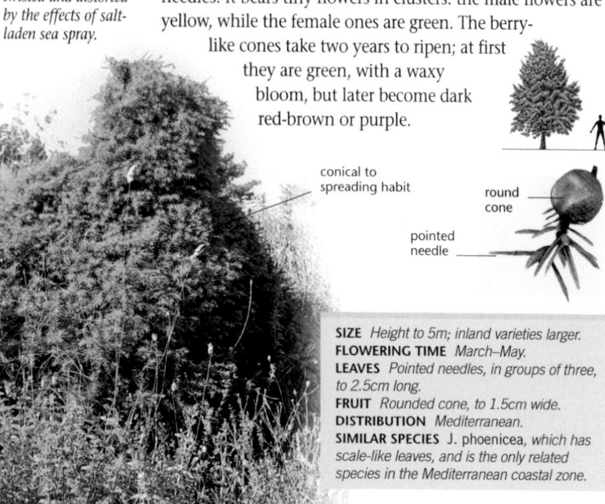

SIZE *Height to 5m; inland varieties larger.*
FLOWERING TIME *March–May.*
LEAVES *Pointed needles, in groups of three, to 2.5cm long.*
FRUIT *Rounded cone, to 1.5cm wide.*
DISTRIBUTION *Mediterranean.*
SIMILAR SPECIES J. phoenicea, *which has scale-like leaves, and is the only related species in the Mediterranean coastal zone.*

Joint-pine

Ephedra distachya (Ephedraceae)

SPREADS *over sand dunes, especially in damp slacks; also found in sandy habitats inland.*

With its rigid, green stems and spreading habit, Joint-pine resembles the prostrate, maritime varieties of Broom (*Cytisus scoparius*), which are found in parts of northwest Europe. However, its flowers are tiny and inconspicuous in contrast to the large yellow pea-flowers of Broom, and its late summer show of bright red berries is both dramatic and unmistakable. With its leaves reduced to scales, perhaps to conserve moisture, the stems of Joint-pine take on the major role of photosynthesis.

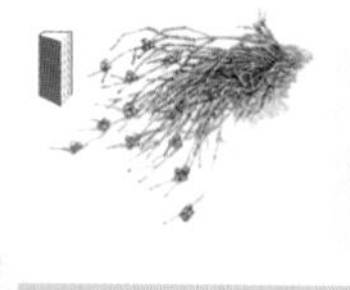

SIZE *Height to 50cm; tiny flowers in clusters to 4mm wide.*
FLOWERING TIME *January–June.*
LEAVES *Tiny and green, on rush-like stems.*
FRUIT *Red berry, 6–7mm wide, in fleshy scales.*
DISTRIBUTION *Western Mediterranean, Atlantic (Portugal).*
SIMILAR SPECIES *None in coastal habitats.*

Stone Pine

Pinus pinea (Pinaceae)

Also known as the Umbrella Pine, because of the very distinctive, low, rounded form of its canopy in open growth, Stone Pine is found along much of the Mediterranean coastline. Its needles are long and rigid, deeply grooved, and grey-green. The rounded, glossy brown cones are borne on recurved shoots, ripening in their third autumn.

GROWS *in sandy as well as rocky coastal habitats, extending a little way inland in similar conditions; frequently planted.*

broad, spreading habit

rounded, umbrella-shaped crown

SIZE *Height and spread to 20m.*
FLOWERING TIME *March–May.*
LEAVES *Needles 8–20cm long, to 2mm wide.*
FRUIT *Edible seeds, in cones to 8cm long.*
DISTRIBUTION *Mediterranean and Portugal.*
SIMILAR SPECIES P. halepensis *and* P. brutia, *which are more irregular in shape and have shorter needles but longer, tapering cones.*

Maritime Pine

Pinus pinaster (Pinaceae)

Conical when young, older Maritime Pines tend to develop a long, bare trunk with a rounded crown. The needles of this evergreen are the longest of any European pine, and its tapering cones are also large and long-lived, often remaining on the tree for several years. It is extensively planted in south and west Europe for commercial forestry and for stabilizing light soil.

PREFERS *lime-rich or sandy soils in the lowlands; often planted to stabilize sand dunes.*

SIZE *Height to 30m or more, spread to 10m.*
FLOWERING TIME *March–April.*
LEAVES *Needles very long, to 20cm.*
FRUIT *Nut-like seed, in cones to 22cm long.*
DISTRIBUTION *Mediterranean, native only in the west; often planted further north and east.*
SIMILAR SPECIES P. nigra *ssp.* laricio, *which has shorter, rounded cones and slightly shorter needles; often planted for dune stabilization.*

Lower plants

The term "lower plant" is often used prejudicially to suggest that such plants are in some way inferior. However, the lower plant groups of algae (seaweeds), mosses, ferns, and allies, together with fungi and lichens, are as well-fitted to their environment as any other species. Indeed, it could be argued that they are better adapted, as they have been around longer – the term "lower" simply relates to the fact that they appeared early on the evolutionary scene. On the coast, most lower plants are found in the more terrestrial habitats, above the high tide line, apart from the seaweeds, which often dominate the intertidal realm.

SANDHILL SCREW-MOSS

SUNBURST LICHEN

SEA SPLEENWORT

DULSE

Sandhill Screw-moss

Syntrichia ruraliformis (Brachytheciaceae)

Few mosses grow in drought-prone areas such as sand dunes as they are unable to regulate water loss like higher plants. However, Sandhill Screw-moss can survive desiccation for extended periods, and then rehydrate itself quickly when it rains. Its ability to grow upwards through fresh sand deposits also adapts it to its habitat.

FAVOURS *sand dunes, especially unstable dunes with bare sand; also shingle; avoids strongly acidic sites.*

SIZE *Height to 4cm.*
FORM *Acrocarpous.*
SPORE PRODUCTION *Occasional; capsules narrowly cylindrical, curved.*
SPORE SEASON *Spring–Summer.*
DISTRIBUTION *Throughout in suitable habitats.*
SIMILAR SPECIES *None on sand dunes. At times considered a dune variety of* S. ruralis.

Whitish Feather-moss

Brachythecium albicans (Brachytheciaceae)

A pleurocarpous moss, with creeping growth and largely horizontal branching, Whitish Feather-moss is a frequent component of low vegetation on sand dunes. The pale, silvery tips of its shoots lend a distinctive gloss to the turf. Examined under a lens, the leaves can be seen to overlap closely, creating a "string-like" appearance. They are triangular in shape, with a fine tip and several lengthwise pleats.

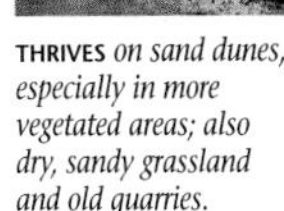

THRIVES *on sand dunes, especially in more vegetated areas; also dry, sandy grassland and old quarries.*

SIZE *Creeping shoots to 8cm long; 2cm thick, depending on the vegetation.*
FORM *Pleurocarpous.*
SPORE PRODUCTION *Rare; ovoid capsule with a conical cap.*
SPORE SEASON *Autumn–Winter.*
DISTRIBUTION *Throughout the region.*
SIMILAR SPECIES *Several pleurocarpous mosses in more vegetated, damper conditions.*

Maidenhair Fern

Adiantum capillus-veneris (Pteridophyta)

DROOPS *from cliffs and rocky clefts; coastal in the north, but also inland in the warmer regions of Europe.*

An elegant, usually drooping, plant of rocky places, the Maidenhair Fern has leaves which are divided into broad, fan-shaped segments. In summer, the margins curl over, protecting the developing sori (spore-bodies), of which there may be up to ten per leaf segment. Although preferring warm localities, it is often to be found in dense shade, especially with water dripping from above. Maidenhair Fern has a remarkably wide global distribution; it is native in Europe, Africa, and both North and South America.

wiry leaf stalk

fan-shaped leaf segments

sori on recurved leaf margins

green, translucent leaves

SIZE *Length of leaves to 40cm.*
LEAVES *Translucent; twice-pinnate with distinctive fan-shaped lobes to 1.5cm long; variably toothed or notched; produced March–October.*
SPORE PRODUCTION *Spores ripen May–September.*
DISTRIBUTION *Mediterranean; Atlantic, north to western England.*
SIMILAR SPECIES *No similar ferns.* Thalictrum *species have similar leaves, and may also be found on the coast, especially* T. minus, *which is found on sand dunes and has similar shaped leaflets, but they do not bear spores and their stalks are green.*

NOTE

Maidenhair Fern is a warmth-loving species, and therefore favours coastal locations in the north of it range. It is also widely cultivated, and has spread onto walls in towns outside its natural range, taking advantage of the locally warmer conditions in the urban "heat island".

Sea Spleenwort

Asplenium marinum (Pteridophyta)

A tufted fern, often growing in crevices towards the top of sea-cliffs, the Sea Spleenwort has bright green leaves, a colour which is obvious even when it grows in deep shade. The leaves are leathery, an adaptation which enables the plant to have green leaves all year round, in spite of being buffeted by winter storms and salt spray.

INHABITS *crevices on sea cliffs and coastal rocks, sometimes extending a little way inland; tolerant of salt spray.*

SIZE *Leaves to 30cm long, sometimes more.*
LEAVES *Bright green, leathery, pinnate.*
SPORE PRODUCTION *June–September.*
DISTRIBUTION *Mediterranean (local); Atlantic, north to southern Norway.*
SIMILAR SPECIES A. obovatum, *which has pinnately divided leaflets; the* Polypodium *ferns, which are simply pinnate, but with unstalked leaflets.*

Great Horsetail

Equisetum telmateia (Pteridophyta)

A robust, patch-forming perennial of damp places, the Great Horsetail is the largest living species of this prehistoric group of plants, which once dominated the earth. It has distinctive fertile and sterile shoots. The unbranched, white fertile shoots, topped with a spore cone, usually emerge in spring before the more persistent, "bottlebrush-like" sterile summer shoots.

THRIVES *in seepage zones and spring-lines on soft cliffs; inland in damp areas, especially on clay soil.*

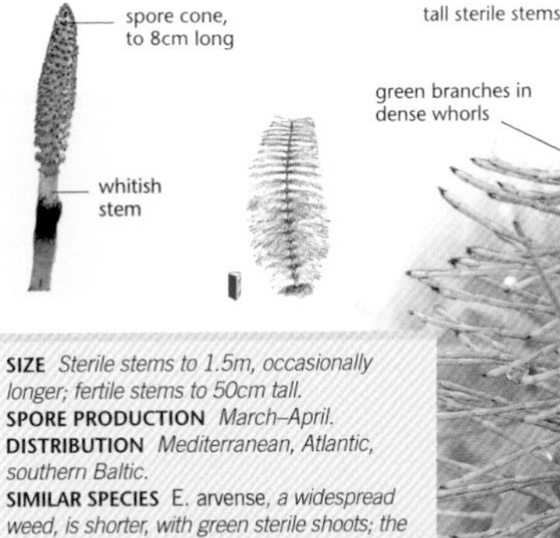

SIZE *Sterile stems to 1.5m, occasionally longer; fertile stems to 50cm tall.*
SPORE PRODUCTION *March–April.*
DISTRIBUTION *Mediterranean, Atlantic, southern Baltic.*
SIMILAR SPECIES E. arvense, *a widespread weed, is shorter, with green sterile shoots; the diminutive* E. variegatum *lacks branches, and creeps around dune slacks.*

Sand Stinkhorn

Phallus hadriani (Fungi)

FOUND *in small groups on sand dunes, usually associated with Marram-grass; also on light soils inland in the south.*

Spending much of the summer as a half-buried, egg-like structure, Sand Stinkhorn has a pinkish skin (volva) that ruptures when the time is right, its fruiting body emerging in a matter of hours. At first, the conical head is covered with a slimy spore mass, which is soon removed by flies. The Sand Stinkhorn smells pleasantly sweet, similar to the scent of hyacinths.

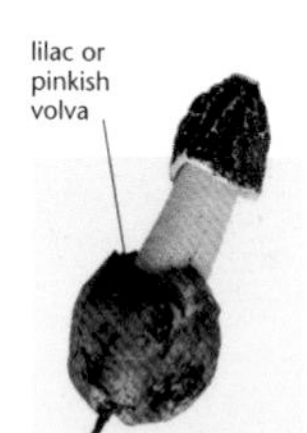

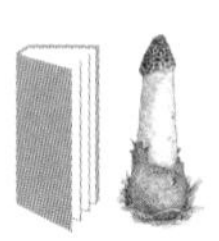

SIZE *Height of fruiting body to 15cm; "egg" to 6cm wide.*
FRUITING PERIOD *Summer–autumn.*
DISTRIBUTION *Throughout the region, but exclusively coastal only in the northern part of its range.*
SIMILAR SPECIES P. impudicus, *which has a white volva and a strong, offensive smell, and is found in woodland, including dune plantations.*

Black Earth-tongue

Geoglossum cookeanum (Fungi)

OCCURS *in mossy turf on sand dunes, and in other freely draining grassland, both coastal and inland.*

Usually growing in small groups in short turf, Black Earth-tongue has a club-shaped, matt black fruiting body on a slender stalk, and is one of a group of fungi that may be found in such nutient-poor places in autumn. Its distinctive shape and colour make it easily recognizable among a variety of greenish earth tongues, the branched coral-fungi, and the often orange, red, or pink waxcap toadstools that characterize such areas.

club-shaped, often flattened fruiting body

matt black flesh

SIZE *Height to 7cm.*
FRUITING PERIOD *July–November.*
DISTRIBUTION *Throughout the region.*
SIMILAR SPECIES *Several other similar black or greenish earth-tongues, some of which require microscopic examination of spores for accurate identification, such as* G. glutinosum, *which has a slimy black fruiting body.*

Reindeer Lichen

Cladonia arbuscula (Lichens)

Several species of the *Cladonia* lichen are found on acidic coastal sands, where the sand has been sufficiently stabilized by the binding effect of plant roots. The grey lichens often become a dominant visual feature, leading to the development of "grey dunes". Although the different species of the genus are very similar, *C. arbuscula* can usually be identified by its yellowish grey appearance and branch tips, all of which curve in one direction.

GROWS *on vegetated sand dunes, especially where the sand is acidic; also acid heathland and peaty areas inland.*

robust, yellowish grey stems

SIZE *Height to 8cm.*
FORM *Fruticose.*
REPRODUCTION *Spores produced in tiny brown structures at the branch tips.*
DISTRIBUTION *Throughout the region.*
SIMILAR SPECIES Cladonia *is a very large and complex genus, with many similar species. Perhaps the most similar are* C. portentosa, *in which the branch tips curve in all directions, and* C. mediterranea, *a southern species, which avoids acidic sites.*

NOTE

The Reindeer Lichen is an example of a fruticose lichen, which grows like a tiny shrub. The other main lichen growth forms are foliose (flat with leafy lobes), such as Xanthoria *(p.76); and crustose (flat, but stuck tightly to a rock), such as* Verrucaria *(p.77).*

Sunburst Lichen

Xanthoria parietina (Lichens)

OCCUPIES *rocky shores, above the black* Verrucaria *zone; also on stable shingle and coastal shrubs.*

The most common yellow-orange lichen, the Sunburst Lichen is characteristically, though not exclusively, maritime. It is tolerant of both salt spray and nitrogen enrichment, and so is often best developed around seabird perches. It also contributes to the characteristic yellow colour zone upwards from the high-water mark on rocks, and is frequently seen on the trunks of high-level salt marsh shrubs such as Shrubby Sea-blite (p.61).

SIZE *Patches to around 10cm wide; often merge together.*
FORM *Foliose.*
REPRODUCTION *Orange fruiting discs, mostly in the centre of the thallus.*
DISTRIBUTION *Throughout the region.*
SIMILAR SPECIES *Several* Caloplaca *species, including* C. marina, *which are maritime and orange, but crustose, lacking raised leafy lobes.*

Black Shields

Tephromela atra (Lichens)

THRIVES *in the splash zone on hard cliffs; also on rocks and walls inland.*

As with the Sunburst Lichen (above), Black Shields (also known as *Lecanora atra*) is found both on rocks in the splash zone and around seabird perches. It is far less noticeable because of its grey colour, but its beauty is apparent up close. It has a pale grey crust, rough and cracking in places to show the underlying rock, and is studded with black fruiting discs, to 3mm wide, each with a pale, raised rim.

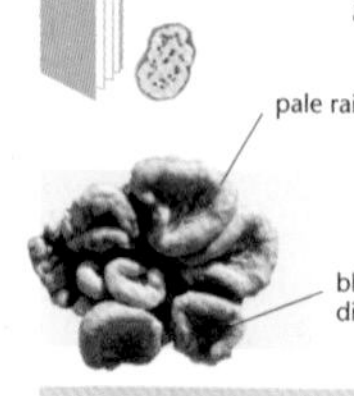

SIZE *Patches to 10cm wide; often merging.*
FORM *Crustose.*
REPRODUCTION *Black fruiting discs, mostly in the centre of the thallus.*
DISTRIBUTION *Throughout the region.*
SIMILAR SPECIES Lecanora gangaleoides, *which is darker, with more regular discs, and* L. rupicola, *in which the discs are usually pinkish, but black when parasitised by another lichen.*

Sea Ivory

Ramalina siliquosa (Lichens)

A conspicuous, greenish grey, bushy lichen, Sea Ivory is often a dominant feature of seaward-facing rocks on the Atlantic coastline. Its stems may be erect or drooping, but they are always brittle. It grows as a series of strap-like stems arising from a common base, which often have pale grey fruiting discs towards their tips. The range of the species extends some way inland, demonstrating the extent of the maritime influence.

INHABITS *sea cliffs and coastal rocks, upwards from the splash zone to rocks several kilometres inland.*

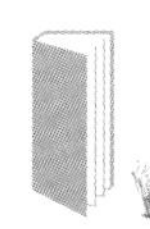

bush-like habit

SIZE *Stems to 7cm long.*
FORM *Fruticose.*
REPRODUCTION *Pale fruiting discs toward the end of the stems.*
DISTRIBUTION *Atlantic, Baltic.*
SIMILAR SPECIES *Several other* Ramalina *species, including* R. calicaris *on coastal tree bark; also* Roccella *species, which have a similar form, but a rubbery texture.*

Tar-lichen

Verrucaria maura (Lichens)

Like a patch of beached oil, the Tar-lichen forms a smooth crust on seashore rocks, so thin that it moulds itself to irregularities on the rock surface. It also bears inconspicuous fruiting bodies. It is a major component of the black zone of coastal rocks, at around the high-tide mark, below the yellow *Xanthoria* zone (see left); this includes several other encrusting *Verrucaria* species, as well as the minutely tufted *Lichina* species, less than a centimetre high.

DOMINATES *rocks and cliff-bases at around the high-water mark of exposed rocky shores.*

forms dull black patches on rocks

often cracked into regular patterns

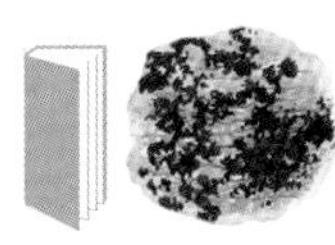

SIZE *Patches often 30cm or more wide.*
FORM *Crustose.*
REPRODUCTION *Small, black fruiting bodies.*
DISTRIBUTION *Atlantic, Baltic.*
SIMILAR SPECIES V. adriatica, *which is found in the Mediterranean;* V. amphibia, *which is glossy and greenish;* V. mucosa, *which is thicker and dark green;* Lichina *species, which are small, tufted black lichens.*

Furbelows

Saccorhiza polyschides (Phaeophyceae)

ATTACHES *firmly to rocks at low-water mark, and into the sublittoral; tolerant of wave attack and current scour.*

Despite its considerable size, Furbelows is an annual plant; it has a very high growth rate in favourable conditions, to 15cm a day at the peak. Its broad, flat, golden-brown fronds are divided into numerous strap-like lobes, although they are less divided in more sheltered conditions. The form of the blade is comparable with several of the *Laminaria* species. Furbelows is, however, easily recognized by its short, wavy-edged stem, and its hemispherical holdfast that is covered in warty protuberances and fringed with small rootlets.

NOTE

As large plants subject to immense wave action, all kelps have substantial holdfasts. Whether root- or disc-like, the holdfasts provide a shelter in the rocky shore ecosystem, in which invertebrates and small fish can take refuge.

SIZE *To 4.5m long.*
GROWTH *Annual.*
REPRODUCTION *Spores produced from the stem and base, October–April.*
DISTRIBUTION *Western Mediterranean, Atlantic.*
SIMILAR SPECIES Laminaria digitata *and* L. hyperborea, *which also have strap-like frond divisions; however, both of them have branching, root-like holdfasts. These two species are differentiated by their stems, which are smooth in* L. digitata *and rough in* L. hyperborea.

Sea Belt

Laminaria saccharina (Phaeophyceae)

A long, belt-like seaweed, Sea Belt is brown, often with a distinct olive tinge. Although quite leathery, it is not very tolerant of exposed conditions, preferring a degree of shelter from the full force of the waves. Dried fronds are good indicators of humidity, when they turn limp and soft, earning this seaweed the other name of Poor Man's Weatherglass.

GROWS *in deep pools and around the low tide line, usually on sheltered rocky shores.*

wavy edge to blade

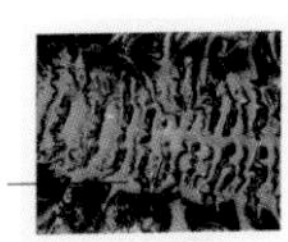

crinkled centre of blade

SIZE *To 4m long.*
GROWTH *Perennial.*
REPRODUCTION *All year, but especially in autumn and winter.*
DISTRIBUTION *Atlantic; the smaller* L. rodriguezi *replaces it in the Mediterranean.*
SIMILAR SPECIES Alaria esculenta *has a distinct midrib within a more flimsy, transparent blade, which often becomes shredded.*

Knotted Wrack

Ascophyllum nodosum (Phaeophyceae)

Sometimes known as Egg Wrack, Knotted Wrack has single, egg-shaped, gas-filled flotation bladders that are found at intervals along the strap-like frond. It is often found on the mid-shore brown seaweed zone, especially where it is sheltered. In exposed conditions, it survives, but is reduced to a tattered tuft of stem bases.

OCCURS *on rocky shores, preferring sheltered conditions, extending into estuaries, although still usually attached to stones.*

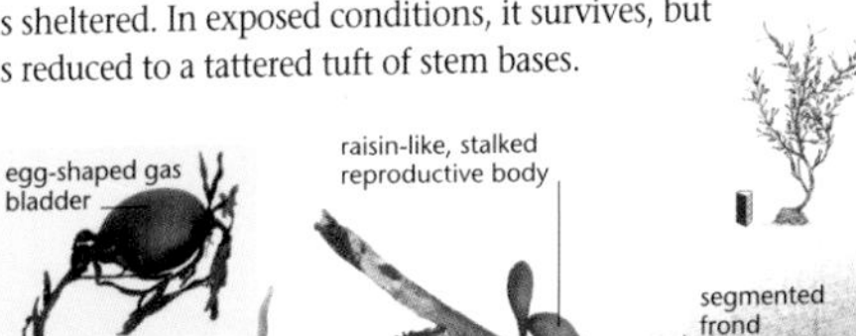

egg-shaped gas bladder

strap-like frond

raisin-like, stalked reproductive body

segmented frond

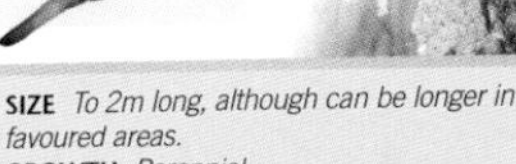

SIZE *To 2m long, although can be longer in favoured areas.*
GROWTH *Perennial.*
REPRODUCTION *April–June.*
DISTRIBUTION *Atlantic.*
SIMILAR SPECIES *None. Small, battered examples could be mistaken for Channelled Wrack (p.81), but the stalked reproductive bodies of Knotted Wrack are distinctive.*

Bladder Wrack

Fucus vesiculosus (Phaeophyceae)

A common mid-shore seaweed, often dominating large areas, the Bladder Wrack has distinctive pairs of gas-filled bladders on its fronds, on either side of the prominent midrib. The bladders help the frond to reach sunlight at high tide, when light penetration is limited by sediment stirred up by waves. In season, the tips of the fronds develop swollen, warty, forked reproductive bodies.

OCCURS *in the middle intertidal zone of rocky shores; also in estuaries and brackish water.*

PERENNIAL

SIZE *Length to 1.5m; shorter with increased exposure to wave action.*
REPRODUCTION *Mid-winter to late summer.*
DISTRIBUTION *Atlantic, Baltic.*
SIMILAR SPECIES *On exposed shores, a short, bladderless form occurs, which is similar to* F. ceranoides *and* F. virsoides, *but these are found in sheltered, low-salinity water and Mediterranean waters, respectively.*

Toothed Wrack

Fucus serratus (Phaeophyceae)

This distinctive seaweed is flattened and generally olive-green. The male plants, however, assume a golden colour when reproducing from the swollen reproductive frond tips. After reproduction the fertile fronds are shed, which coupled with damage from storms, leads to a lower dominance of this species over winter. The fronds are often covered by the white spiral tubes of the worm *Spirorbis spirorbis*.

INHABITS *the lower-middle zone of sheltered rocky shores; dominant below the Bladder Wrack zone.*

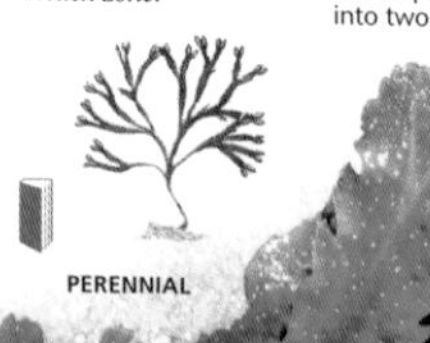

PERENNIAL

frond split into two

prominent midrib

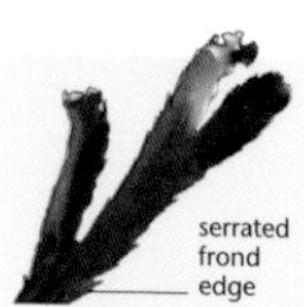

SIZE *Length to 70cm, or more, up to 2m, in very sheltered conditions; frond width about 2cm.*
REPRODUCTION *May–October, peaking in late summer.*
DISTRIBUTION *Atlantic; western Baltic.*
SIMILAR SPECIES *None – the serrated frond margins distinguish it from the bladderless* F. ceranoides, F. spiralis, *and* F. virsoides.

Flat Wrack

Fucus spiralis (Phaeophyceae)

Lacking bladders or serrations, Flat Wrack is one of the more tricky wracks to identify. The tendency of its fronds to twist is a useful clue, but this characteristic is far from constant, and its reproductive bodies may have to be examined to confirm identification. A key feature is the prominent midrib on the flat, spiralling frond.

FAVOURS *sheltered to moderately exposed rocky shore habitats, at mid- to upper-shore levels; can also penetrate estuaries.*

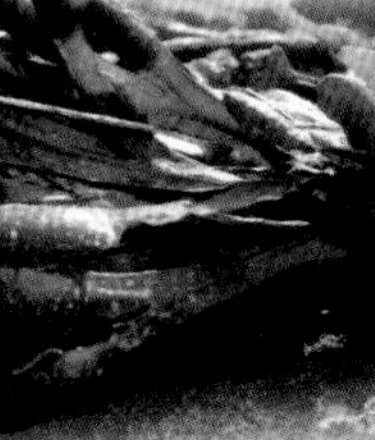

SIZE *Length to 40cm.*
REPRODUCTION *July–September.*
DISTRIBUTION *Atlantic.*
SIMILAR SPECIES F. ceranoides, *which does not spiral and has more regular branching, with often rather pointed reproductive bodies;* F. virsoides, *which has fan-like branching, more spherical reproductive bodies, and is restricted to the Mediterranean.*

rounded reproductive tip

PERENNIAL

Channelled Wrack

Pelvetia canaliculata (Phaeophyceae)

An upper-shore brown seaweed, Channelled Wrack has tough fronds that lack a midrib and are curled over lengthwise to create a channel, perhaps to help conserve water. The colour is yellow when wet and dark brown to almost black when dry, which the plant often is at its usual location on the shore. Channelled Wrack is highly resistant to desiccation, capable of tolerating 65 per cent water loss. In fact, permanent immersion is sufficient to kill it.

FOUND *on upper rocky shores, from above the* Fucus *species up to the splash zone.*

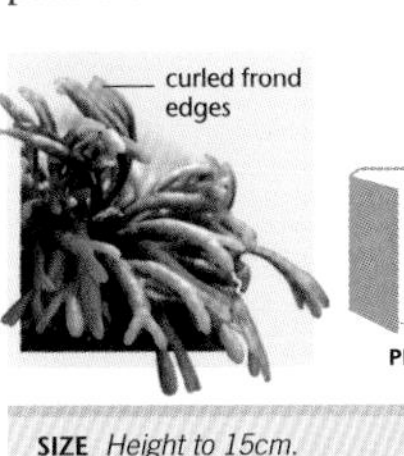

PERENNIAL

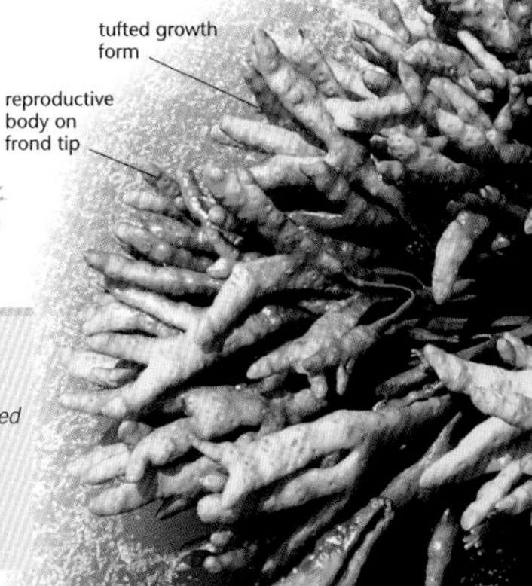

SIZE *Height to 15cm.*
REPRODUCTION *August–September.*
DISTRIBUTION *Atlantic.*
SIMILAR SPECIES *No other brown seaweed occurs so high on the shore. The lichen* Lichina pygmaea *shares the same zone, and though it is only 1.5cm tall, it has a superficial resemblance to a stunted Channelled Wrack.*

THRIVES *in sheltered, often estuarine waters, at the low-water mark and below; sometimes loosely attached to rocks, but mostly free-floating.*

BASE PERENNIAL

Japweed

Sargassum muticum (Phaeophyceae)

Introduced accidentally from the Pacific with commercial oysters, Japweed, or Wireweed, has adapted well to life in the Atlantic. Too well in fact, as it is spreading rapidly and becoming a serious pest species, one for which control is proving difficult. This is largely because of its extremely high reproductive capability, its rapid growth, and the ability of free-floating specimens to stay alive and to drift in the sea currents. This seaweed has a distinctive growth form. It is long and bushy and has regular alternate branches, which are clothed in oval, leaf-like blades. Its fronds bear numerous small, spherical gas-filled bladders.

SIZE *Length typically to 1m, but often to 10m in favoured spots.*
REPRODUCTION *Summer.*
DISTRIBUTION *Mediterranean; Atlantic, as far north as southern Norway.*
SIMILAR SPECIES S. vulgare, *a native species, which has larger, lance-shaped, leaf-like side branches;* S. hornschuchi, *which has side branches with wavy margins, and is found in the Mediterranean;* Halidrys siliquosa, *which is orange in colour, and has numerous segmented, elongated gas bladders.*

NOTE

This introduced species is spreading rapidly, leading to ecological impacts on other seaweeds and marine plants such as Eel-grass (p.54). Human activites are also affected – the dense beds of Japweed inhibit navigation.

Thongweed

Himanthalia elongata (Phaeophyceae)

The long, strap-like structures which give Thongweed its name are actually the reproductive bodies, which turn from olive-green to yellowish when ripe, before being shed during autumn. The perennial part of the plant is a mushroom-shaped disc, about 3cm wide, attached to rocks by a short stalk.

ATTACHES *to stones and rocks on the lower rocky shore, often abundant in deep rock pools; tolerates moderate exposure.*

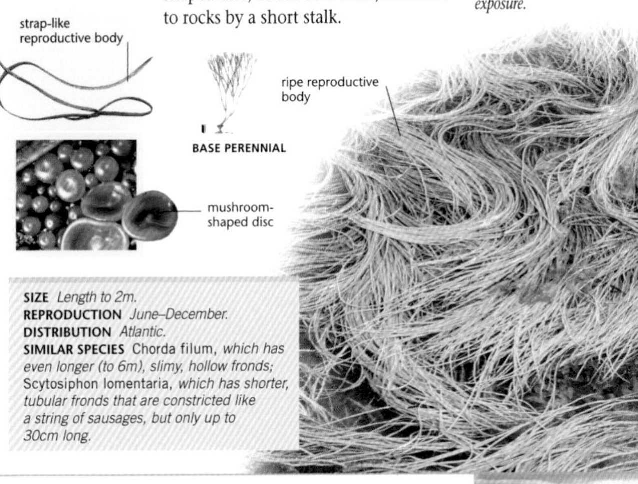

SIZE *Length to 2m.*
REPRODUCTION *June–December.*
DISTRIBUTION *Atlantic.*
SIMILAR SPECIES Chorda filum, *which has even longer (to 6m), slimy, hollow fronds;* Scytosiphon lomentaria, *which has shorter, tubular fronds that are constricted like a string of sausages, but only up to 30cm long.*

Peacock's Tail

Padina pavonica (Phaeophyceae)

With its fan-like fronds curved into a funnel shape, Peacock's Tail is a very unusual brown seaweed. Its fronds are somewhat stiffened with chalky deposits, and its outer surface is distinctively banded brown and green, with concentric lines of fine hair. The inside of the funnel is usually a lime-green colour. A representative of a largely tropical group of seaweeds, in Europe it grows best in the warmer waters of the Mediterranean.

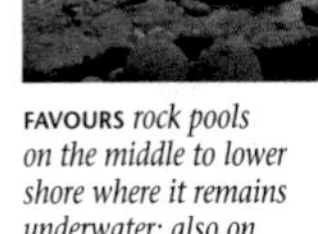

FAVOURS *rock pools on the middle to lower shore where it remains underwater; also on rocks at the lowest tidal levels.*

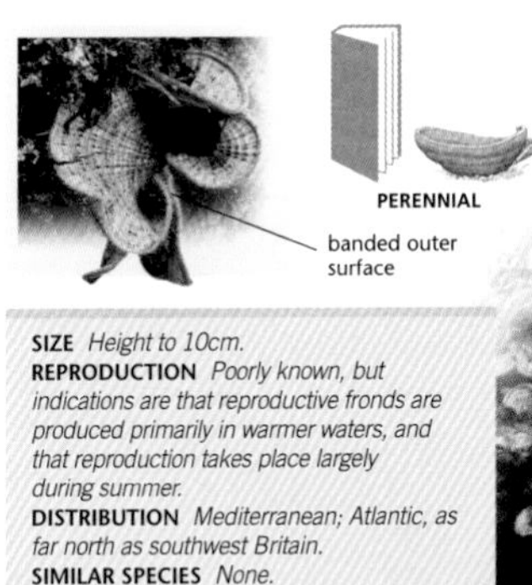

SIZE *Height to 10cm.*
REPRODUCTION *Poorly known, but indications are that reproductive fronds are produced primarily in warmer waters, and that reproduction takes place largely during summer.*
DISTRIBUTION *Mediterranean; Atlantic, as far north as southwest Britain.*
SIMILAR SPECIES *None.*

Coral-weed

Corallina officinalis (Rhodophyceae)

THRIVES *along the edges of mid-shore rock pools, and on open lower rocky shores, into the shallow sublittoral.*

An inhabitant of lower rocky shores, Coral-weed derives its name from the chalky deposits on its fronds, which also makes it resilient to grazing and the action of waves. Its pink, encrusting holdfast, similar to Pink Paint (below), is up to 7cm wide, and is anchored to rocks, shells, or larger brown seaweeds. The regular, opposite-branching pattern gives Coral-weed a feather-like appearance. The colour of this species varies according to age and light conditions.

pinkish purple colour

SIZE *Height to 12cm.*
FORM *Perennial, dying in parts in winter.*
REPRODUCTION *All year; urn-shaped reproductive bodies near the tips of the fronds.*
DISTRIBUTION *Atlantic, Baltic.*
SIMILAR SPECIES C. elongata, *which is more densely tufted;* Lomentaria articulata, *which is segmented, but not calcified;* Jania rubens, *which has a two-branched pattern.*

Pink Paint

Lithothamnion sp. (Rhodophyceae)

ENCRUSTS *rocks and stones at lower shore levels; also forms loose reefs below the low tide mark.*

A group of closely related encrusting, calcified red algal species, Pink Paint forms characteristic reddish, often zoned, splashes on rocks at the lower intertidal levels. Because the algae are so small, identification to species level is very difficult for the non-specialist, but most are within the genera *Lithothamnion* and *Lithophyllum*.

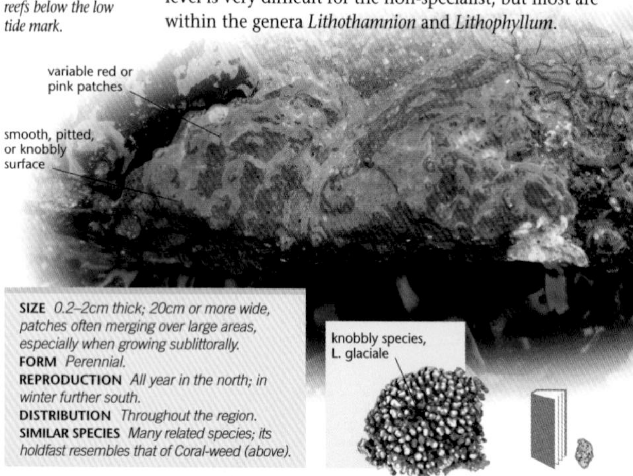

SIZE *0.2–2cm thick; 20cm or more wide, patches often merging over large areas, especially when growing sublittorally.*
FORM *Perennial.*
REPRODUCTION *All year in the north; in winter further south.*
DISTRIBUTION *Throughout the region.*
SIMILAR SPECIES *Many related species; its holdfast resembles that of Coral-weed (above).*

Pepper Dulse

Osmundea pinnatifida (Rhodophyceae)

Formerly known as *Laurencia pinnatifida*, Pepper Dulse has tough, flattened fronds, which resist wave attack. It is often, therefore, a dominant species on exposed coasts. At mid-shore levels, it is usually yellow-brown, but lower down the shore, in reduced light conditions, it is brownish purple. Its common name is derived from its pungent taste; it is sometimes dried and used as a spice.

FOUND *on rocky shores at mid- to lower levels; forming an extensive turf on moderately exposed coasts.*

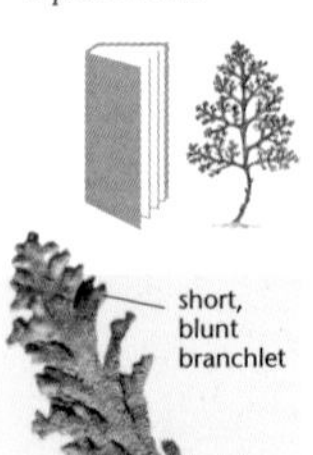

flattened frond

alternate branches

SIZE *Length to 20cm, but much shorter, typically to 8cm, at mid-shore levels.*
FORM *Perennial.*
REPRODUCTION *Spring.*
DISTRIBUTION *Mediterranean (rare); Atlantic.*
SIMILAR SPECIES *Largely replaced by* O. truncata *in the Mediterranean;* Laurencia obtusa, *which has a similar branching pattern, but the fronds are cylindrical and rose-pink.*

Carragheen

Chondrus crispus (Rhodophyceae)

With flat fronds, dividing up to five times, Carragheen, or Irish Moss, is shaped like a fan. However, its shape, size, and colour varies according to its habitat: smaller plants with narrower lobes are found in more exposed conditions; in strong sunlight, it turns green in colour; and underwater, the tips of the fronds are often iridescent. Carragheen is collected as a source of carrageenan, a thickening agent used in ice-cream and other foods.

OCCURS *on rocky shores, at lower intertidal levels and in rock pools; tolerant of low salinity.*

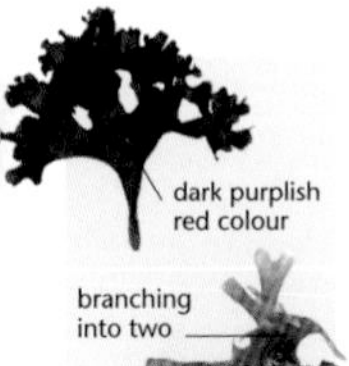

branching into two

SIZE *Length to 20cm.*
FORM *Perennial; individual fronds die after 2–3 years, but new ones regrow from the long-lived holdfast.*
REPRODUCTION *Autumn–Spring.*
DISTRIBUTION *Mediterranean, Atlantic, western Baltic.*
SIMILAR SPECIES Mastocarpus stellatus *has curved branch margins, forming a gutter.*

Dulse

Palmaria palmata (Rhodophyceae)

INHABITS *lower rocky shores, usually attached to stones or stems of kelps; extends into the sublittoral.*

A leathery red seaweed of the lower shore kelp zone, Dulse has broad annual fronds arising from a perennial disc-shaped holdfast. The fronds are usually palmately divided into flat lobes, although in sheltered, muddy conditions, the frond may be divided into narrow, linear lobes. An important food in parts of its range, Dulse is eaten fresh, dried, or fried, and is a good source of protein and valuable dietary supplements.

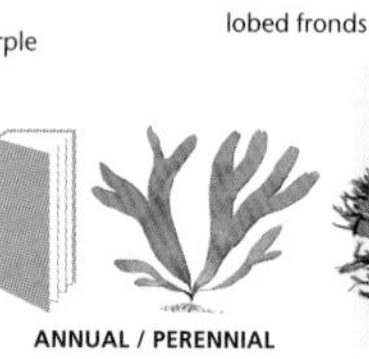

SIZE *Length to 50cm, sometimes longer.*
REPRODUCTION *All year.*
DISTRIBUTION *Atlantic, Baltic.*
SIMILAR SPECIES *Red Rags (below), which is more leathery and unbranched;* Halarachnion ligulatum *(Mediterranean), which is pink, more regularly branched, with a gelatinous texture;* Calliblepharis ciliata, *which has a branched holdfast, and numerous marginal leaflets.*

Red Rags

Dilsea carnosa (Rhodophyceae)

ATTACHES *to rocks at around the extreme low-water mark, down into the sublittoral.*

Sometimes confused with Dulse (above), but unpalatable, Red Rags is found at the lowest tide levels, in deep rock pools, or among stands of kelp. The very leathery, red-brown fronds are an oval shape, arising from a narrow, cylindrical stem and holdfast; older fronds are often split at the tip, but are never palmately lobed, like Dulse.

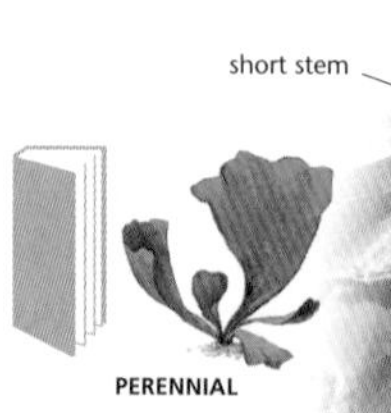

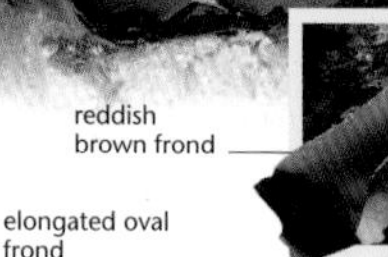

SIZE *Length to 30cm; width to 20cm.*
REPRODUCTION *December–January.*
DISTRIBUTION *Atlantic, Baltic.*
SIMILAR SPECIES *Dulse, which is thinner and more slippery;* Halarachnion ligulatum *(Mediterranean), which is more regularly branched;* Calliblepharis ciliata, *which has a branched holdfast;* Gelidium latifolium, *which has narrower, ribbon-like main fronds.*

Purple Laver

Porphyra umbilicalis (Rhodophyceae)

Although a red seaweed, Purple Laver is green when young, purplish red as it matures, and black when it dies. Its frond is a thin, but tough, flat membrane that is attached to a rock in the centre, forming the umbilicus (hence its scientific name). Purple Laver's ability to cope with exposure to the air allows it to exploit a wide range of intertidal conditions.

FOUND *attached to rocks in sandy habitats, throughout intertidal zone; resists desiccation and wave action.*

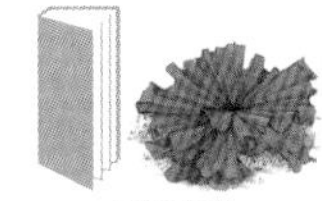

PERENNIAL

flat, membranous frond

SIZE *Width to 20cm.*
REPRODUCTION *Vegetative and sexual; all year.*
DISTRIBUTION *Throughout the region.*
SIMILAR SPECIES *None, although according to experts, based upon molecular investigations, this widespread seaweed should be treated as a complex of very closely related species.*

Red Sea-oak

Phycodrys rubens (Rhodophyceae)

Although bearing translucent, flimsy-looking fronds, Red Sea-oak is able to survive in the high energy, lower shore environment, on and among the stems of kelps. Although the frond blades of Red Sea-oak are shredded and removed by winter storms, the tough, wiry midrib, which is capable of regrowing a new blade the following spring, remains.

GROWS *attached to rocks, kelps, and other brown seaweeds in lower-shore rock pools down to the sublittoral.*

pale crimson fronds

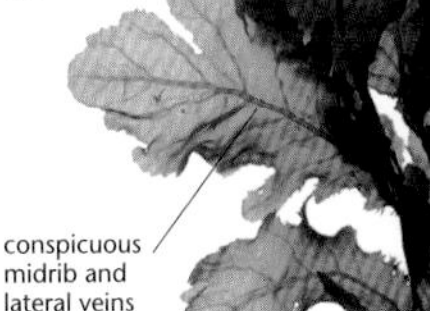

SIZE *Length to 15cm.*
REPRODUCTION *Autumn.*
DISTRIBUTION *Atlantic, southern Baltic.*
SIMILAR SPECIES Delesseria sanguinea, *which has a similar frond texture and epiphytic (growing on or attached to a living plant) habit, but has crimson fronds with wavy margins;* Apoglossum ruscifolium, *which has flattened, branched fronds but lacks secondary ribs.*

PERENNIAL

Sea-lettuce

Ulva lactuca (Ulvophyceae)

TOLERATES *brackish water; found in estuaries and on rocky shores; also free-floating.*

A widespread green seaweed, tolerant of most conditions except extreme exposure, Sea-lettuce is found throughout the intertidal zone. It thrives especially in brackish water, but is also found in shallow inshore waters. Its flat frond is often split or divided, and has a wavy edge. Fertile plants are sometimes recognizable by changes to the marginal colour of the frond, becoming yellowish green in male plants and dark green in the female.

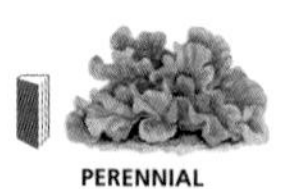

PERENNIAL

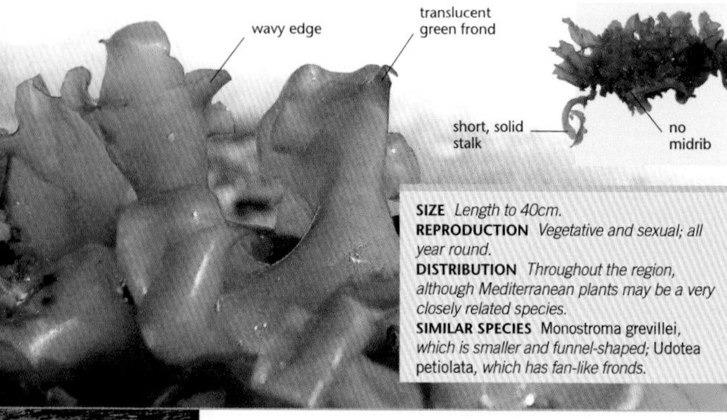

SIZE *Length to 40cm.*
REPRODUCTION *Vegetative and sexual; all year round.*
DISTRIBUTION *Throughout the region, although Mediterranean plants may be a very closely related species.*
SIMILAR SPECIES Monostroma grevillei, *which is smaller and funnel-shaped;* Udotea petiolata, *which has fan-like fronds.*

Mermaid's Cup

Acetabularia acetabulum (Ulvophyceae)

INHABITS *rock pools and lower rocky shores, into the sublittoral; attached to rocks and stones.*

With the appearance of a small mushroom, Mermaid's Cup has fronds that are petal-like and fused to a flat or concave disc on a stalk that is stiffened with chalky deposits. Despite its complex appearance, each plant comprises only a single cell – it is one of the largest unicellular organisms known and, consequently, much used in scientific research.

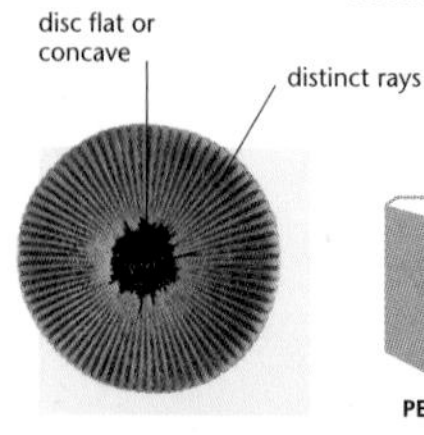

PERENNIAL

long, stiff stalk

greenish white disc

SIZE *Height to 8cm; disc to 1cm wide.*
REPRODUCTION *Year round. The single large nucleus in a root-like rhizoid divides repeatedly; daughter nuclei are transported to the tips of the branches, where they are released.*
DISTRIBUTION *Mediterranean.*
SIMILAR SPECIES *None, but often known by the name* A. mediterranea; *largely a sub-tropical group of species.*

Gut-weed

Ulva intestinalis (Ulvophyceae)

Previously and perhaps more familiarly known as *Enteromorpha intestinalis*, Gut-weed forms an inflated, irregularly constricted, tube-like frond, resembling an intestine. It requires high nutrient levels, and so is very well developed near sewage outfalls or other nutrient-rich discharges from the land. It blankets estuarine mudflats, especially in late summer, and is a favoured food of grazing wildfowl.

ATTACHES *to rocks and stones on sandy or muddy shores; abundant in low salinity areas.*

irregular constrictions

pale to bright green fronds

unbreached, inflated fronds

SIZE *Length to 80cm.*
REPRODUCTION *All year round, but concentrated in the summer in the north; coordinated with the phases of the moon.*
DISTRIBUTION *Throughout the region; has a global distribution.*
SIMILAR SPECIES U. linza, *which has flattened, crinkly fronds;* U. compressa, *which is also flat, often with branched fronds.*

ANNUAL

Rock Cladophora

Cladophora rupestris (Ulvophyceae)

A rigid, dark green, reed-like green seaweed, Rock Cladophora has a distinctive rough texture. The individual filaments are made up of a single column of cells, which arises from the resilient basal plate of rhizoids or root-like attachment structures. In exposed localities, this seaweed may be reduced to a short, almost moss-like turf, even into the splash zone.

FOUND *especially on lower to middle rocky shores; some forms tolerate exposure to the air or low salinity.*

thread-like fronds

PERENNIAL

SIZE *Length to 20cm.*
REPRODUCTION *All year round.*
DISTRIBUTION *Throughout the region.*
SIMILAR SPECIES *Several other* Cladophora *species, especially in the Mediterranean, which have mainly microscopic differences;* Chaetomorpha linum, *which is not tufted, and has filaments that look like a string of beads when magnified.*

dark, often bluish green

Invertebrates

There are tens of thousands of invertebrate species found around the coastlines of Europe. These include many insects, often the dominant group in habitats above the tideline, and numerous intertidal and sublittoral species, which are drawn from an array of invertebrate groups. The selection of species profiled on the following pages are grouped into phyla, and these have been arranged as far as possible such that similar-looking species, irrespective of their relationships, appear close to each other for ease of comparison.

ALCYONIDIUM HIRSUTUM

PEACOCK WORM

VARIEGATED SCALLOP

SALTMARSH MINING-BEE

Yellow Sea-squirt

Ciona intestinalis (Urochordata)

Despite their primitive appearance, sea-squirts are chordates, sharing a common ancestry with the vertebrates. The Yellow Sea-squirt, or Sea-vase, is one of the larger solitary species. The free end bears two siphons, one to take in water (inhalant), and the other to expel it (exhalant), after extracting food and oxygen. Its translucent body allows the internal organs to be seen clearly.

ATTACHES *to rocks and large seaweeds around the low-water mark; often found on harbour walls and in estuaries.*

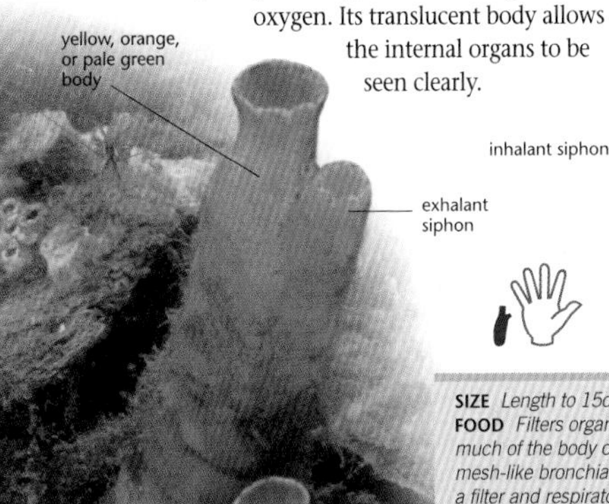

SIZE *Length to 15cm.*
FOOD *Filters organic matter from sea water; much of the body cavity is occupied by a mesh-like bronchial sac, which acts as both a filter and respiratory surface.*
SHORE ZONE *Low-water mark to shallow sublittoral.*
DISTRIBUTION *Mediterranean, Atlantic.*
SIMILAR SPECIES *None.*

Red Sea-squirt

Ascidia mentula (Urochordata)

Attached by its flank to a rock, the Red Sea-squirt is a large, solitary species, although it often clusters with others of the same species in favourable locations. The body is oval, with a thick, rubbery test (skin), which is often colonized by other encrusting marine life. The siphons are inconspicuous, but the inhalant siphon is marked by small white lobes around the rim.

FOUND *on hard substrates at lower intertidal levels; tolerant of reduced salinity.*

red or greenish body
small, white lobes around siphon
thick, tough skin

SIZE *Length to 18cm, occasionally longer.*
FOOD *Filters organic matter from sea water.*
SHORE ZONE *Lower intertidal, down to the continental shelf edge.*
DISTRIBUTION *Throughout the region.*
SIMILAR SPECIES A. conchilega *is smaller and green, with red lobes round the siphons;* Phallusia mamillata *is large and pear-shaped, with smooth rounded bumps.*

White Clathrina

Clathrina coriacea (Porifera)

FORMS *patches on rocks, stones, and especially under overhangs.*

Although it has a soft, delicate texture, the White Clathrina is classified as a calcareous sponge, due to the calcareous spicules which stiffen its body. The shape of the microscopic spicules is three-pointed. On a larger scale, White Clathrina appears as a meshwork of branched and fused thin-walled tubes that are usually white, but can also be grey, pink, orange, or yellow. It does not have any free, erect branches.

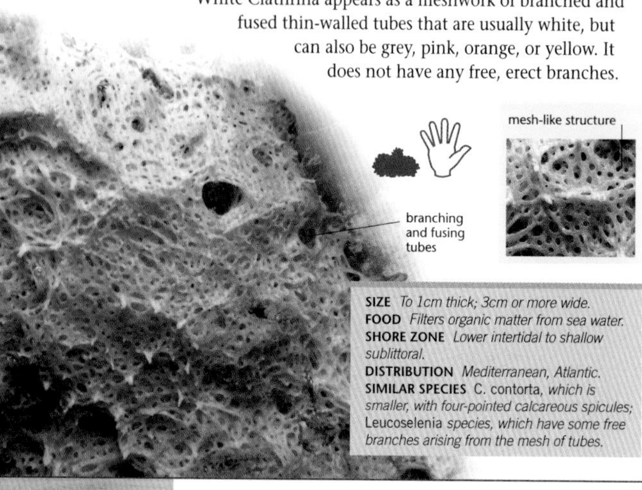

SIZE *To 1cm thick; 3cm or more wide.*
FOOD *Filters organic matter from sea water.*
SHORE ZONE *Lower intertidal to shallow sublittoral.*
DISTRIBUTION *Mediterranean, Atlantic.*
SIMILAR SPECIES C. contorta, *which is smaller, with four-pointed calcareous spicules;* Leucoselenia *species, which have some free branches arising from the mesh of tubes.*

Rough Scallop Sponge

Myxilla incrustans (Porifera)

ENCRUSTS *rock surfaces, stone, and shells in the lower intertidal zone of rocky shores, often in areas exposed to current or wave action.*

The most common yellow sponge in sites exposed to waves or currents, the Rough Scallop Sponge consists of a thick spreading cushion, with raised ridges. Usually pale to sulphur-yellow in colour, the surface consists of numerous deep, labyrinthine channels, across which run cobweb-like strands of tissue. Large pores (oscula) are usually clearly visible, often on raised cones or in rows on surface ridges.

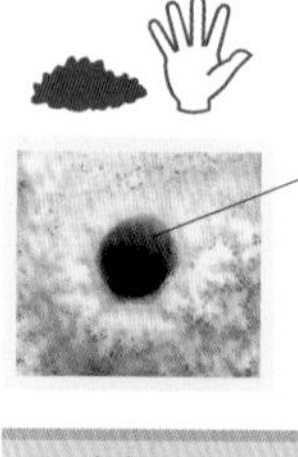

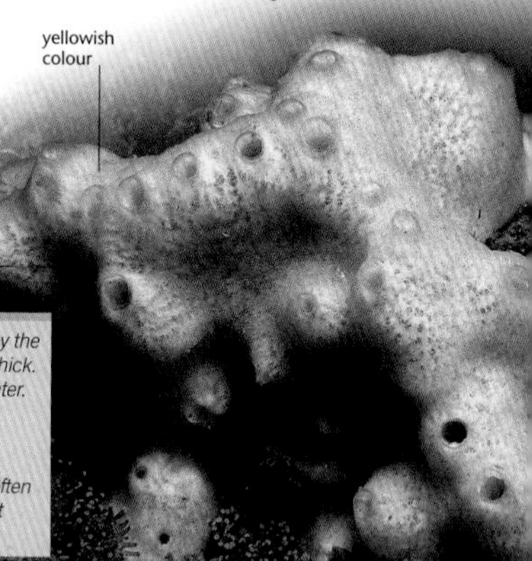

SIZE *Indeterminate spread, limited only by the size of its substrate; several centimetres thick.*
FOOD *Filters organic matter from sea water.*
SHORE ZONE *Lower intertidal to deep sublittoral.*
DISTRIBUTION *Mediterranean, Atlantic.*
SIMILAR SPECIES M. rosacea, *which is often reddish or orange, and may develop erect branches from the encrustation.*

Breadcrumb Sponge

Halichondria panicea (Porifera)

So called because of its crumb-like texture, the Breadcrumb Sponge can grow into massive reefs. Its mounds incorporate sand and stones, holdfasts, and other detritus, and provide niches for a wide range of marine animals. Orange and yellow are two frequent colour forms, but a green form containing symbiotic algae is characteristic of well-lit conditions.

FORMS *patches and mounds on rocks and seaweed stems, and reefs, especially in areas with strong currents.*

irregular patches and mounds

green colour form

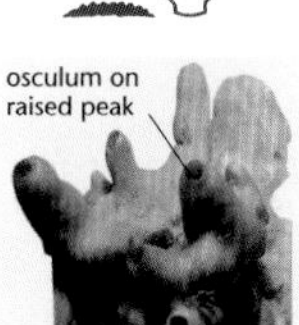

SIZE *To 20cm wide and 10cm thick.*
FOOD *Filters organic material from sea water.*
SHORE ZONE *Lower intertidal to shallow sublittoral.*
DISTRIBUTION *Mediterranean, Atlantic, southern Baltic.*
SIMILAR SPECIES H. bowerbanki, *which is paler, more inclined to produce lobes and branches, and lacks large oscula.*

Purse Sponge

Grantia compressa (Porifera)

Usually shaped like a hot-water bottle – a flattened sac with a single terminal opening – the Purse Sponge sometimes adopts different growth forms, from cylindrical to a series of contorted, flat lobes. White or cream forms are most usual, with occasional brown specimens. Such variation leads to identification problems. For example, much remains to be discovered about the presence and identity of purse-shaped sponges in the Mediterranean.

OCCURS *in clumps in brown algae and rocks; often under overhangs.*

SIZE *Height to 4cm, often smaller.*
FOOD *Filters organic material from sea water.*
SHORE ZONE *Lower intertidal to shallow sublittoral.*
DISTRIBUTION *Atlantic, north from northern France.*
SIMILAR SPECIES Sycon ciliatum, *which has a similar shape, but with a hairy surface, and its osculum has a fringe of spines.*

Sea Mat

Membranipora membranacea (Bryozoa)

FORMS *lacy patches on the fronds of kelps, on moderately exposed rocky shores; tolerant of brackish conditions.*

Bryozoa are composed of numerous tiny zooids, each with a crown of tentacles. Together they create distinct shapes, from flat plates, to knobbly masses, to branched, twig-like structures. The Sea Mat is a common encrusting species, forming flat sheets on broad kelp fronds, typically towards the base of the frond, which is less vulnerable to wave damage than the tip.

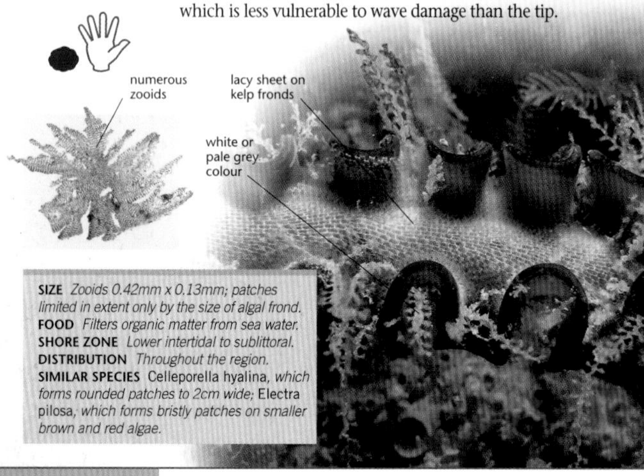

SIZE *Zooids 0.42mm x 0.13mm; patches limited in extent only by the size of algal frond.*
FOOD *Filters organic matter from sea water.*
SHORE ZONE *Lower intertidal to sublittoral.*
DISTRIBUTION *Throughout the region.*
SIMILAR SPECIES Celleporella hyalina, *which forms rounded patches to 2cm wide;* Electra pilosa, *which forms bristly patches on smaller brown and red algae.*

Alcyonidium hirsutum

Alcyonidium hirsutum (Bryozoa)

GROWS *in patches on the fronds of brown and red seaweeds at middle to lower levels on rocky shores.*

Several uncalcified, rubbery *Alcyonidium* species form encrustations on the fronds of brown and red seaweeds, especially the *Fucus* species. *A. hirsutum* is distinctly velvety to the touch, due to the presence of zooids of different sizes and shapes. It often develops erect, fleshy projections, and then resembles *A. diaphanum*, which is a lobe-forming species, often washed up as Sea-chervil (p.211) on beaches.

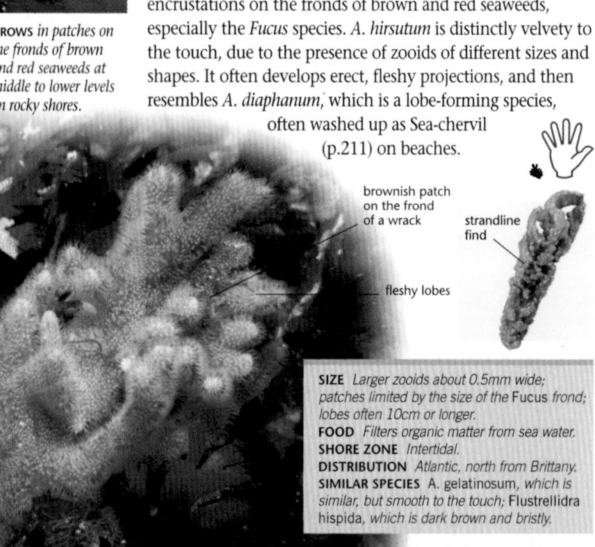

SIZE *Larger zooids about 0.5mm wide; patches limited by the size of the* Fucus *frond; lobes often 10cm or longer.*
FOOD *Filters organic matter from sea water.*
SHORE ZONE *Intertidal.*
DISTRIBUTION *Atlantic, north from Brittany.*
SIMILAR SPECIES A. gelatinosum, *which is similar, but smooth to the touch;* Flustrellidra hispida, *which is dark brown and bristly.*

Sea-fir

Abietinaria abietina (Cnidaria)

There are many colonial hydrozoa which show an erect, branched form similar to the Sea-fir, although the zig-zag stem and alternate branching pattern of this species are fairly distinctive. Arranged regularly along both stem and branches, again in an alternating manner, are numerous flask-shaped structures, called hydrothecae. These house the individual animals that make up the colony.

ATTACHES *to shells, stones, and other hard structures around the lower intertidal levels of a rocky shore.*

zig-zag stem

alternate branches

SIZE *Colonies up to 20cm long.*
FOOD *Filters organic material from sea water.*
SHORE ZONE *Low-water mark to shallow sublittoral.*
DISTRIBUTION *Mediterranean, Atlantic.*
SIMILAR SPECIES Sertularella gay *has less regular branching and more strictly alternate hydrothecae;* Obelia geniculata *has just one hydrotheca at each node of the zig-zag stem.*

White-weed

Sertularia cupressina (Cnidaria)

With a robust, bushy appearance that lends itself to flower-arrangment and other ornamental purposes, White-weed is commercially harvested by dredging. It is tolerant of both sand scour and inundation, and thus forms a distinctive component of the fauna of channel edges, often with the similar but less bushy *Hydrallmania falcata*. Its white stems, with feathery branchlets, are quite distinctive.

FOUND *around and below extreme low-water mark in sheltered, sandy areas.*

feathery appearance

SIZE *Colonies to 50cm long.*
FOOD *Filters organic material from sea water.*
SHORE ZONE *Extreme low-water mark to shallow sublittoral.*
DISTRIBUTION *Atlantic.*
SIMILAR SPECIES Thuiaria thuja *has a more dense, cylindrical shape;* S. argentea *is generally smaller, but otherwise identical, and perhaps not a truly different species.*

Scarlet and Gold Star-coral

Balanophyllia regia (Cnidaria)

ATTACHES *to rocks and stones in low shore pools and gullies on moderately exposed rocky coasts.*

Closely related to tropical reef-building corals, the Scarlet and Gold Star-coral has a calcified, spongy, cup-shaped skeleton (corallum). While reef-forming species are colonial, the Scarlet and Gold Star-coral is solitary, though individuals aggregate in favourable areas.

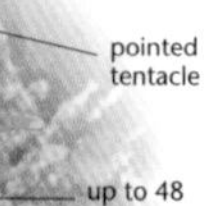

SIZE *Height to 8mm; 1cm wide; spread to 2.5cm when tentacles are out.*
FOOD *Small invertebrates and plankton.*
SHORE ZONE *Extreme low-water mark to shallow sublittoral.*
DISTRIBUTION *Mediterranean; Atlantic, as far north as southern Britain.*
SIMILAR SPECIES Leptopsammia pruvoti *is taller, with more tentacles;* B. italica *is larger.*

Dead-man's Fingers

Alcyonium digitatum (Cnidaria)

LIVES *under rocks and in pools of rocky shores, often covering large areas sublittorally.*

Lacking a skeleton, but stiffened with calcareous spicules, Dead-man's Fingers is a soft coral with broad, fleshy lobes. A very long-lived species, it prefers areas with strong wave turbulence or currents. Spawning occurs in winter, so that the settlement and commencement of feeding can coincide with the plankton boom in spring.

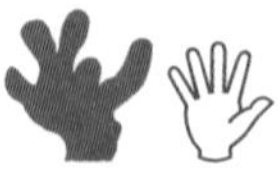

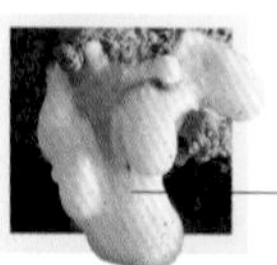

SIZE *Lobe-like colonies to 25cm long; encrustations to 10cm deep.*
FOOD *Filters plankton from the sea.*
SHORE ZONE *Lower intertidal and shallow sublittoral.*
DISTRIBUTION *Atlantic.*
SIMILAR SPECIES A. palmatum *and* A. glomeratum, *which are both more slender, often erect, and redder.*

Starlet Sea-anemone

Nematostella vectensis (Cnidaria)

A tiny anemone, with a column rarely more than 1.5cm high, the Starlet Sea-anemone is translucent, although the tentacles show up pale and starry against a muddy background. The tentacles are relatively long, between 9 and 18 in number, and are very strongly adhesive. They occur in two whorls, the outer being longer than the inner. The Starlet Sea-anemone is usually found in great abundance – densities of more than 10,000 per square metre have been recorded in places. It is found widely on the east and west coasts of north America, but in Europe, it is restricted to the southeast coast of Britain.

FOUND *in mud or on vegetation in brackish lagoons at or above high water; occasionally in open mudflats near fresh-water inflows.*

tentacle with faint bars

outer tentacles longer than inner ones

SIZE *Height to 1.5cm, some individuals to 6cm, their growth perhaps related to abundant food supply.*
FOOD *Preys on small snails and insect larvae.*
SHORE ZONE *Supratidal lagoons and mid-tidal shores.*
DISTRIBUTION *Atlantic (south and east Britain), very localized and possibly overlooked.*
SIMILAR SPECIES Edwardsia ivelli, *which is a tiny, worm-like anemone to 2cm long, with 12 transparent tentacles, and burrows in soft mud in just one saline lagoon in Sussex, but may now be extinct.*

NOTE

The Starlet Sea-anemone shares its habitat with other rare invertebrates, such as Tenellia adspersa *and* Gammarus insensibilis *– this accords its habitat the highest level of protection under European law.*

Beadlet Anemone

Actinia equina (Cnidaria)

INHABITS *all rocky shore habitats, from exposed to sheltered, especially favouring rock pools; tolerant of temperature and salinity fluctuations.*

Named after the ring of blue beads, called acrorhagi, at the top of its column, the Beadlet Anemone is very variable in colour, with red, green, brown, and orange forms. It is the most characteristic intertidal anemone, and when the tide recedes, it remains visible as a blob of jelly on a rock, with its tentacles fully retracted. This anemone species is very territorial, displaying aggressive behaviour when it comes into contact with the tentacles of adjacent anemones. It attacks the victim with its stinging cells, so that the stung anemone either crawls away or drops off the rock.

NOTE

Beadlet Anemones have long been regarded as variable in colour. However, the spotted form has recently been treated as a separate species, A. fragacea. *The green, lower shore form could also receive species status as* A. prasina.

SIZE *Height to 5cm, fully extended; basal width to 5cm.*
FOOD *Preys on invertebrates and small fish.*
SHORE ZONE *Upper intertidal pools to the shallow sublittoral.*
DISTRIBUTION *Mediterranean, Atlantic.*
SIMILAR SPECIES A. fragacea, *which is a little larger, red with green or yellowish spots, and is usually found in the lower intertidal zone; Dahlia Anemone (p.99), which is often red, but much larger, and has a warty column;* A. cari, *a greenish Mediterranean species, which is noticeably conical when its tentacles are retracted.*

Snakelocks Anemone

Anemonia viridis (Cnidaria)

ATTACHES *to rocks or large algae, in full light, at the lower intertidal levels of rocky shores.*

Very distinctive, with its mass of sinuous tentacles, the Snakelocks Anemone rarely retracts its tentacles fully, even when the tide is out. It has a brownish column, but the tentacles are mostly bright green (as they contain symbiotic algae) with purple tips. The tentacles feel very sticky to the touch, and can cause a painful rash on tender skin.

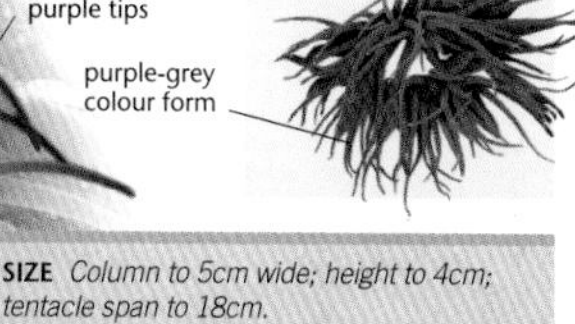

SIZE *Column to 5cm wide; height to 4cm; tentacle span to 18cm.*
FOOD *Carnivorous diet of fish, invertebrates.*
SHORE ZONE *Lower intertidal and very shallow sublittoral.*
DISTRIBUTION *Mediterranean; Atlantic, as far north as southern Norway.*
SIMILAR SPECIES Bunodactis verrucosa, *which has a longer, very warty column.*

Dahlia Anemone

Urticina felina (Cnidaria)

FOUND *in pools and gulleys on rocky shores, downshore from the neap low-water mark; tolerates low salinity.*

Sometimes forming dense carpets, the Dahlia Anemone is a large species, with up to 160 short, stout tentacles. It is extremely variable in colour. It may be a single colour or a combination of white, yellow, orange, red, blue, grey, purple, and brown, although forms with some red are most frequent. The tentacles are usually banded, and the column has numerous warts to which gravel and shell fragments stick.

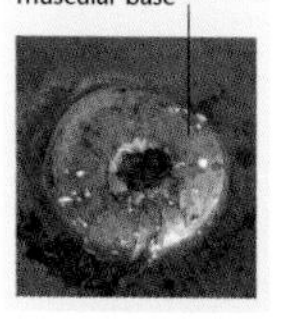

SIZE *Column base to 15cm wide; maximum tentacle span of 20cm.*
FOOD *Crabs, shrimps, molluscs, and worms.*
SHORE ZONE *Lower intertidal to deep sublittoral.*
DISTRIBUTION *Atlantic, north from Biscay; southern Baltic.*
SIMILAR SPECIES U. eques, *which is much bigger (to 30cm wide), and is less warty.*

Sagartia elegans

Sagartia elegans (Cnidaria)

INHABITS *rock pools, often under overhangs, at middle to lower shore levels on rocky coasts; also sublittoral.*

This small sea-anemone occurs in a range of colour forms. Its column is flared at both ends, the top bearing up to 200 tentacles. The column bears prominent suckers, and small dark pores that house defensive threads called acontia. This sea-anemone is attached tightly to rocks by means of a broad basal disc. It reproduces asexually, by budding from the base.

SIZE *Height to 8cm; column to 3cm wide; spread to 5cm.*
FOOD *Small invertebrates.*
SHORE ZONE *Middle and lower intertidal pools to shallow sublittoral.*
DISTRIBUTION *Mediterranean, Atlantic.*
SIMILAR SPECIES S. troglodytes *and Daisy Anemone (below), which also have suckers on the column, but often with debris attached.*

Daisy Anemone

Cereus pedunculatus (Cnidaria)

ANCHORS *in crevices in rock pools; on sandy coasts and small rocks buried in the sand, with only the disc showing.*

With a trumpet-shaped column that is widest at the top, the Daisy Anemone has a distinctive shape, especially with its crown of numerous, short tentacles. Usually brownish to purple, there are colour forms with variegated discs and tentacles in yellow, red, and black. Often, only the disc is visible as the animal lodges its column in a crevice or buries it in sand.

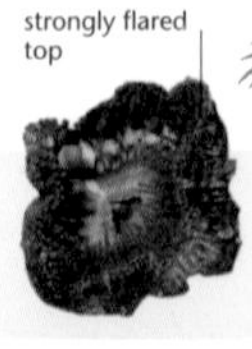

SIZE *Height and spread to 15cm.*
FOOD *Small invertebrates and fish.*
SHORE ZONE *Mid-intertidal to the shallow sublittoral.*
DISTRIBUTION *Mediterranean; Atlantic, north to Britain.*
SIMILAR SPECIES Sagartia troglodytes *is usually smaller, and the column is broadest at the base; it also has fewer, longer tentacles.*

Bootlace Worm

HIDES *coiled in crevices and under stones, on the lower zones of rocky shores, often among kelp holdfasts.*

Lineus longissimus (Nemertea)

The nemerteans, or ribbon-worms, are long, slimy, unsegmented, worm-like animals, unrelated to the true worms. While some are very small, others such as the Bootlace Worm, can grow up to 10m long. An active hunter, it subdues its prey with the cluster of sticky filaments at the end of its snout. Its head has deep, sensory lateral grooves, and 10–20 eye-spots on either side.

slimy, often striped body

SIZE *Often up to 10m long; occasionally to 30m; width to 5mm.*
FOOD *Marine invertebrates and carrion.*
SHORE ZONE *Mid-shore, down to the shallow sublittoral.*
DISTRIBUTION *Atlantic, Baltic.*
SIMILAR SPECIES L. bilineatus, *which is shorter, lacks eye-spots, and has two pale stripes running down the length of its body.*

Symsagittifera roscoffensis

FOUND *in pools on sandy beaches in sunlit conditions, ideal for its symbiotic alga* Tetraselmis convolutae.

Symsagittifera roscoffensis (Platyhelminthes)

Formerly known as *Convoluta roscoffensis*, this green flatworm leaves green trails on the sand at low tide due to the presence of symbiotic algae in its tissues. These algae photosynthesize food for the flatworm, so much so that the adult flatworm has lost functional feeding parts – they are truly solar-powered animals.

little differentiation of body parts

green due to presence of algae

SIZE *Typically up to 1.5mm long, but can grow to 5mm or more.*
FOOD *Sugars from its associated algae.*
SHORE ZONE *Middle to lower intertidal.*
DISTRIBUTION *Atlantic, northern France and southern Britain.*
SIMILAR SPECIES Convoluta convoluta, *which is often larger, lives around seaweeds, and retains a functional mouth.*

Honeycomb Worm

Sabellaria alveolata (Annelida)

FORMS *reefs on rocks on the lower intertidal zone, and in the shallow sublittoral.*

An ideal habitat of the Honeycomb Worm has rocks within a sandy matrix. The free-swimming larvae settle within a few months of hatching, and cement sand grains and shell fragments into a tube in which they live. The tube is anchored to rocks, and often thousands of tubes aggregate together to form massive, humped reefs. These reefs are resilient enough to withstand strong wave action. The Honeycomb Worm has feeding tentacles around the mouth, upward-pointing gills on the back, and a tube-like, forward-pointing tail section. However, the worm itself is unlikely to be seen, when covered by water, the tentacles can be seen protruding from the tubes.

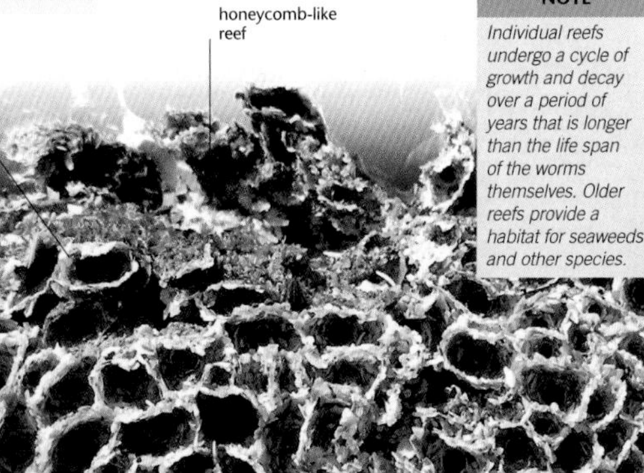

NOTE

Individual reefs undergo a cycle of growth and decay over a period of years that is longer than the life span of the worms themselves. Older reefs provide a habitat for seaweeds and other species.

SIZE *Length to 4cm; tube opening 5–8mm wide.*
FOOD *Filters organic matter.*
SHORE ZONE *Lower intertidal, into the shallow sublittoral.*
DISTRIBUTION *Mediterranean; Atlantic, north to Scotland.*
SIMILAR SPECIES S. spinulosa, *which is shorter, and makes tubes that are often solitary on rocks or shells, or if aggregated, never form extensive humped reefs;* Lanice conchilega, *which is a much longer worm, and lives in sandy tubes that protrude from the sand at the low-water mark.*

Peacock Worm

Sabella pavonina (Annelida)

A slender, tube-dwelling fanworm, the Peacock Worm inhabits a flexible, rubbery mud tube, which sticks out of the substrate by up to 10cm. When covered by water, the worm extends its crown of feathery tentacles in order to filter food and sort material for building its tube. It is, however, very wary and withdraws rapidly if it senses movement or a sudden change in light.

FAVOURS *muddy sand flats, at the low-water mark and below, also within seagrass beds.*

SIZE *Length to 30cm; tentacle span to 4cm.*
FOOD *Filters organic matter.*
SHORE ZONE *Lower intertidal to shallow sublittoral.*
DISTRIBUTION *Mediterranean, Atlantic.*
SIMILAR SPECIES S. spallanzanii *has tentacles in a series of nested whorls;* Megalomma vesiculosum *has fewer tentacles, and inhabits tubes projecting 2–3cm from the substrate.*

Keelworm

Pomatoceros lamarcki (Annelida)

Adult tube-worms in the Serpulidae family live their adult life in a hard, chalky tube secreted by the worm. The Keelworm's tube is triangular in cross-section. The worm itself is small and segmented. It has a crown of feeding tentacles, one of which is modified into an operculum – a plug to close the tube entrance when the worm withdraws.

LIVES *in tubes on rocks, stones, and shells, in the intertidal zone and below, on most types of coastline.*

chalky tube on rocks

tube with longitudinal ridges

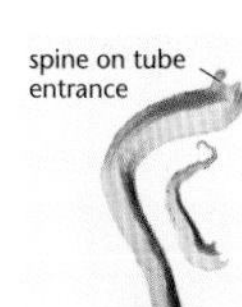

SIZE *Length to 2.5cm; tube 3–5mm wide.*
FOOD *Filters organic debris.*
SHORE ZONE *Middle to lower shore zones, and into the sublittoral.*
DISTRIBUTION *Mediterranean, Atlantic.*
SIMILAR SPECIES Pomatostegus polytrema *has lattice-like tubes;* Serpula vermicularis *has yellow or pink tubes anchored at the base; spirorbid worms have tightly coiled tubes.*

Estuary Ragworm

Hediste diversicolor (Annelida)

BURROWS *to a depth of 20cm in muddy sand, especially in sheltered estuarine habitats; tolerant of low salinity.*

An important bait species for fishing, the Estuary Ragworm is a large, active worm with a hundred or more segments, each with a pair of parapods – fleshy lobes which serve as paddles or legs. The head is furnished with a pair of large palps, four small eyes, two antennae, and four pairs of tentacles. Previously known as *Nereis diversicolor*, there is still much controversy among specialists about its correct name.

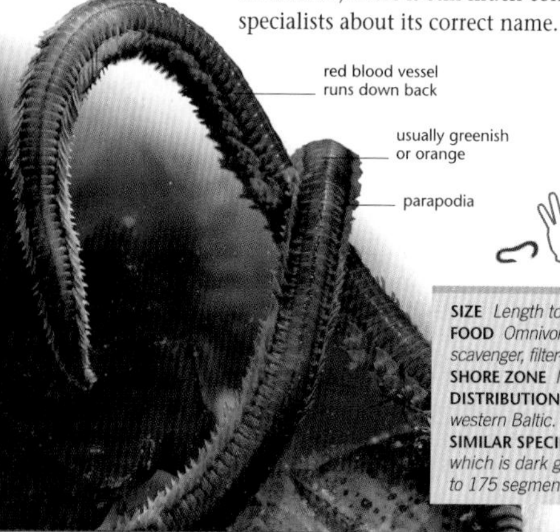

flattened, segmented body

SIZE *Length to 12cm.*
FOOD *Omnivorous, feeding as a carnivore, scavenger, filter-feeder, and deposit-feeder.*
SHORE ZONE *Mid-shore to shallow sublittoral.*
DISTRIBUTION *Mediterranean, Atlantic, western Baltic.*
SIMILAR SPECIES *King Rag* (Nereis virens), *which is dark green, to 40cm tall and has up to 175 segments.*

Green-leaf Worm

Eulalia viridis (Annelida)

FAVOURS *rocks and seaweed-covered areas in the lower intertidal zone; often in kelp holdfasts.*

A member of the paddleworm family, named after their large parapods (flattened lobes), the Green-leaf Worm is one of the largely green species, and an active predator. It is highly segmented, has large paired parapods, two pairs of short antennae, with a fifth antenna between the eyes, and four pairs of tentacles. The Green-leaf Worm is slender, and often very long. One of its most distinctive features is the large, rounded, green egg masses it produces.

SIZE *Length to 15cm.*
FOOD *Carnivorous, feeding on small invertebrates.*
SHORE ZONE *Lower intertidal to shallow sublittoral.*
DISTRIBUTION *Mediterranean, Atlantic.*
SIMILAR SPECIES E. bilineata *is smaller and paler;* Phyllodoce lamelligera *is larger and darker green, with long, pointed paddles.*

Lugworm

Arenicola marina (Annelida)

Sometimes known as the Blow Lug, the Lugworm is a large, fleshy, cylindrical worm that inhabits U-shaped burrows in intertidal mud and sand flats. The worm draws currents of water through its burrow, digesting micro-organisms and organic detritus from the sand in the water. The inlet hole in the sand is marked by a shallow depression, and the outlet by a sandy cast. Lugworms are a major food source for fish and long-billed wading birds such as the Curlew (p.194), and are much valued as bait for angling.

OCCUPIES *tubes in clean or muddy sand flats, at middle to lower beach levels and in estuaries.*

SIZE *Length to 25cm.*
FOOD *Digests organic material from ingested sand.*
SHORE ZONE *Middle tidal levels to shallow sublittoral.*
DISTRIBUTION *Mediterranean, Atlantic, western Baltic.*
SIMILAR SPECIES A. defodiens, *which is often black, makes deeper burrows at lower levels on exposed beaches;* Arenicolides *species, which lack the narrowed "tail", make finer casts, and occupy mud patches between stones;* Scalibregma inflatum, *which has a more strongly inflated, wrinkled forward section.*

NOTE

The Lugworm cast provides a visual clue to the presence of the worm to potential predators. Therefore, they try to minimize surface waste by retaining their faeces in their tail segments. Nevertheless, the tail of the Lugworm is often nipped off by predatory fish and worms.

Cotton-spinner

Holothuria forskali (Echinodermata)

OCCUPIES *rocky shores at the lowest intertidal zones, where there are deposits of soft sediment for feeding.*

One of the larger sea-cucumbers, Cotton-spinner is so called because it ejects sticky white threads from its rear end when alarmed. It lacks the 5-radiate symmetry of other echinoderms, but has the typical feature of numerous tube feet on the underside. It is dark and coarse, with white-tipped projections contrasting against a dark skin.

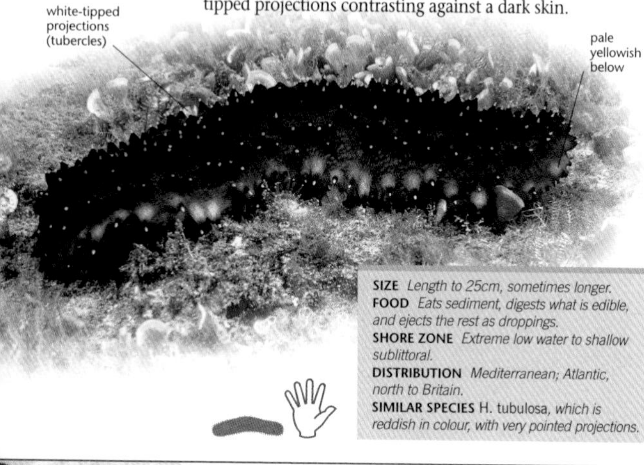

SIZE *Length to 25cm, sometimes longer.*
FOOD *Eats sediment, digests what is edible, and ejects the rest as droppings.*
SHORE ZONE *Extreme low water to shallow sublittoral.*
DISTRIBUTION *Mediterranean; Atlantic, north to Britain.*
SIMILAR SPECIES H. tubulosa, *which is reddish in colour, with very pointed projections.*

Northern Sea-cucumber

Cucumaria frondosa (Echinodermata)

OCCURS *among rocks, stones, and kelp holdfasts at the lower levels of rocky shores.*

A large, fat sea-cucumber, the Northern Sea-cucumber has enormous bushy tentacles. It is also known as Sea-pudding, although this name is used for its various tropical relatives too. Its skin is blackish or very dark brown, but the tentacles and mouth region can be quite colourful, with areas of white or red.

blackish or coloured tentacles

10 long, highly branched tentacles

SIZE *Length to 50cm.*
FOOD *Deposit- and filter-feeder.*
SHORE ZONE *Lower intertidal to deep sublittoral.*
DISTRIBUTION *Mediterranean, Atlantic.*
SIMILAR SPECIES *Cotton-spinner (above) is the only other large black sea-cucumber likely to be found, but it lacks the large bushy tentacles of the Northern Sea-cucumber.*

Serpent-star

Ophiura ophiura (Echinodermata)

With short, rigid arms (only up to four times the diameter of the disc), the Serpent-star is an example of one of the two main forms of brittle-star. It is less brightly coloured than related species – a necessary camouflage due to its surface-dwelling habit, as opposed to sheltering in crevices.

FOUND *on or shallowly buried in muddy sands and gravel, at and around the extreme low-water mark.*

row of short spines at base of each arm

disc covered with dense scales

relatively large disc

sandy or brownish camouflage colour

SIZE *Disc to 3.5cm wide; arm span to 25cm.*
FOOD *Omnivorous; scavenges, hunts, and filter-feeds.*
SHORE ZONE *Extreme low water to sublittoral.*
DISTRIBUTION *Mediterranean, Atlantic.*
SIMILAR SPECIES O. albida, *which is reddish and smaller, its disc only to 1.5cm wide;* Ophiocomina nigra, *which is larger, darker, and fringed with dense spines.*

Slender Brittle-star

Amphiura filiformis (Echinodermata)

One of the brittle-stars that have very long, flexible, sinuous arms, the Slender Brittle-star is a long-lived species, believed to have a lifespan of 20 years or more. When buried in the sediment, its arms extend vertically for up to 4cm into the water current to filter-feed. It is often found alongside *A. chiajei*, but the arms of this brittle-star rest on the surface to feed on deposited organic material.

LIVES *on or buried in fine muddy sand, at and below the extreme low-water mark.*

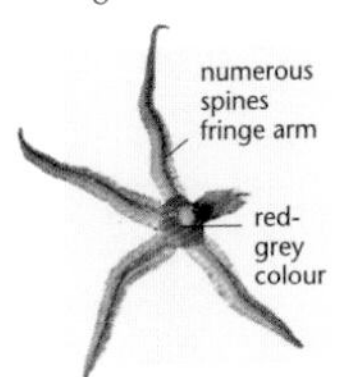

SIZE *Disc to 1cm wide; arm span to 22cm.*
FOOD *Plankton and detritus, filtered from flowing water.*
SHORE ZONE *Extreme lower intertidal to sublittoral.*
DISTRIBUTION *Mediterranean, Atlantic.*
SIMILAR SPECIES A. chiajei *(see above);* Acrocnida brachiata, *which is darker, and filters food with its arm tips.*

Common Starfish

Asterias rubens (Echinodermata)

INHABITS *intertidal and sublittoral areas on sand, gravel, or rock; also penetrates into estuaries and low salinity areas.*

One of the largest and most familiar starfish in Europe, the Common Starfish is found in almost any lower intertidal or sublittoral habitat, sometimes in huge swarms. Although it can be large, in exposed conditions or where the food supply is poor, most are less than 20cm wide. The smaller specimens are rather stiff, but the larger ones can be quite floppy. The arms have a row of blunt, white spines down the mid-line, together with many pale granules that are arranged less regularly. The skin is normally orange-brown, but this is sometimes masked by blue or purple tinges.

5 broad, fleshy arms

orange-brown skin

upper side with granules and blunt spines

NOTE

A voracious predator, the Common Starfish tackles bivalves, such as mussels, by pulling apart the shells with the suckers of its numerous tube-feet. It then everts its stomach and digests the prey externally.

SIZE *Diameter to 50cm.*
FOOD *Preys on bivalve molluscs, worms, crustaceans, and other echinoderms; also scavenges carrion.*
SHORE ZONE *Intertidal to deep sublittoral.*
DISTRIBUTION *Atlantic, Baltic.*
SIMILAR SPECIES Marthasterias glacialis, *which is even larger, with distinct rows of broad-based, sharp white spines running down the arms, and is also found in the Mediterranean;* Leptasterias muelleri, *which is smaller, to 10cm wide, and has a northerly distribution.*

Sand-star

BURIES *itself shallowly in clean sand, at and around the extreme low-water mark.*

Astropecten irregularis (Echinodermata)

One of the stiff starfish, with sharply pointed arms, the Sand-star lives in clean, sandy habitats, either on or just buried underneath the surface. While it has a camouflaging sandy colour, it sometimes has a purple tip at the end of each arm and a similarly coloured spot at the centre of the disc.

SIZE *Width to 20cm.*
FOOD *Preys on bivalves, crustaceans, worms, and brittle-stars.*
SHORE ZONE *Extreme low water to deep sublittoral.*
DISTRIBUTION *Mediterranean, Atlantic.*
SIMILAR SPECIES A. arancianus, *which is larger, to 60cm wide;* Luidia sarsi, *which has longer and less rigid arms.*

Green Starlet

FAVOURS *rock pools and algae-encrusted boulders on rocky shores, even in exposed conditions.*

Asterina phylactica (Echinodermata)

Cushion-stars are almost pentagonal, with very short protruding arms. The Green Starlet or the Small Cushion-star is distinguished by its tiny size and spiny or granular upper surface, with a marked dark green or brown star-pattern. However, it is an enigmatic species, discovered as recently as 1979 as being distinct from one colour form of *A. gibbosa*. As a result, its precise distribution is not fully known.

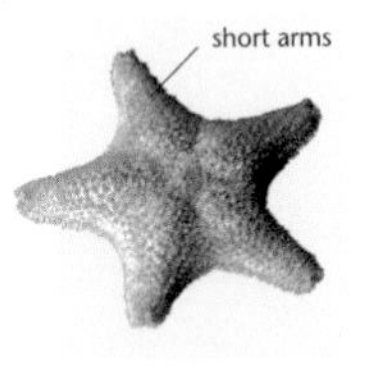

spiny or granular upper surface

SIZE *Diameter to 1.5cm.*
FOOD *Scavenges dead animal and plant remains.*
SHORE ZONE *Intertidal and shallow sublittoral.*
DISTRIBUTION *Mediterranean; Atlantic, north to Ireland.*
SIMILAR SPECIES A. gibbosa, *which is larger, to 5cm wide, and smoother.*

Edible Sea-urchin

Echinus esculentus (Echinodermata)

OCCURS *on rocky shores, at the lowest intertidal levels, where it grazes on algae and other marine organisms.*

A large sea-urchin with a reddish purple globular shell, or test, the Edible Sea-urchin is covered in spines, which are all of a similar length. The spines are pale red with violet tips, and grow from whitish shell projections. In shallower water, the shell may grow with a more flattened profile. It lives for about ten years, although older specimens have been reported on the basis of growth rings on the shell.

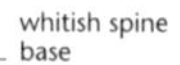

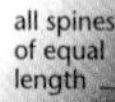

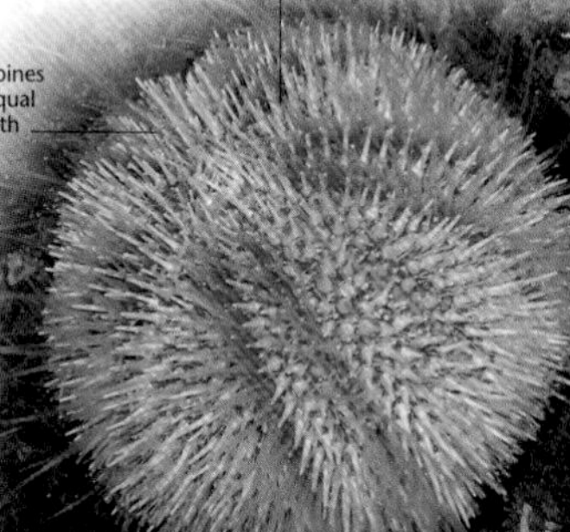

whitish spine base

globular test of dead specimen

SIZE *Width to 18cm; height to 15cm.*
FOOD *Grazes on algae, especially kelps, and encrusting bryozoans and barnacles.*
SHORE ZONE *From the low-water mark to deep water.*
DISTRIBUTION *Atlantic.*
SIMILAR SPECIES E. acutus *has a more conical test and white-tipped spines;* Sphaerechinus granulatus *has white-tipped,*

Green Sea-urchin

Psammechinus miliaris (Echinodermata)

INHABITS *rocky shores and areas with a coarse, sandy bottom, often among seagrass beds.*

One of the smaller members of the sea-urchin group, the Green Sea-urchin has a slightly flattened shell, with five rounded angles that are visible when viewed from above. The spines are short and usually have deep violet tips, which contrast with the greenish shell. Its colour, however, varies with the habitat: those from shallower water tend to be darker and more purplish.

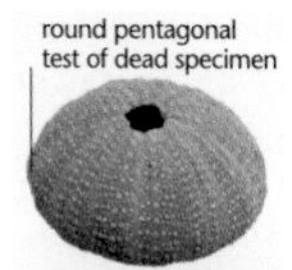

round pentagonal test of dead specimen

spines with violet tips

short spines

round mouth

SIZE *Width to 5cm; height to 4cm.*
FOOD *Grazes young barnacles, young bivalves, sea-squirts, and algae.*
SHORE ZONE *Middle intertidal to sublittoral.*
DISTRIBUTION *Atlantic, western Baltic.*
SIMILAR SPECIES P. microtuberculatus *is smaller, with red-tipped spines;* Paracentrotus lividus *has a more flattened shell and long, dark green or purple spines.*

Black Sea-urchin

Arbacia lixula (Echinodermata)

FAVOURS *rocky coastlines, occupying the lower shore levels into the shallow sublittoral.*

Often an abundant species on Mediterranean rocky coasts, the Black Sea-urchin can be a considerable hazard to swimmers, as its sharp black spines can inflict a painful wound to unprotected skin. It has a low, flattened profile, and a very large mouth, occupying more than half of the lower surface of the shell. After it dies, the spines are worn off to reveal a contrasting pinkish test.

SIZE *Shell to 5cm wide; spines to 3cm long.*
FOOD *Grazes on algae and seagrass beds.*
SHORE ZONE *Lower intertidal to shallow sublittoral.*
DISTRIBUTION *Mediterranean, Atlantic (Portugal).*
SIMILAR SPECIES *Green Sea-Urchin (p.110), which has a greenish, rounded pentagonal shell; often occur together.*

Heart-urchin

Echinocardium cordatum (Echinodermata)

INHABITS *permanent burrows, to a depth of 15cm, in clean sand on sheltered beaches.*

Well-adapted to a burrowing lifestyle, with a streamlined shell and flattened, backward-pointing spines, the Heart-urchin digs through sand very efficiently, using the broad spines on its underside. It is rarely seen, unless excavated from the sediment, into which it can bury itself very rapidly. Much more familiar are the dead shells, having lost their spines, which are known as Sea Potatoes.

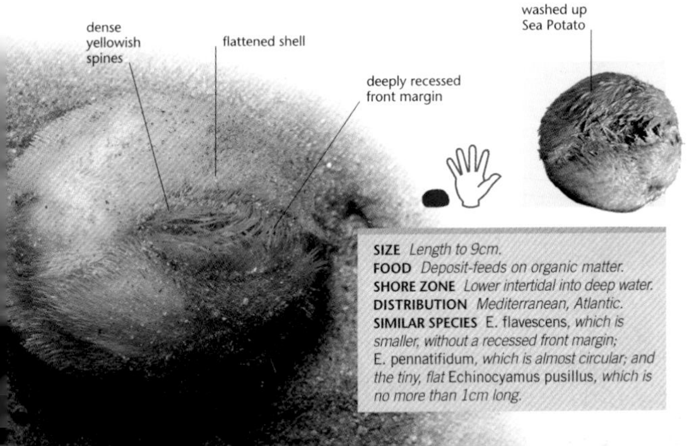

SIZE *Length to 9cm.*
FOOD *Deposit-feeds on organic matter.*
SHORE ZONE *Lower intertidal into deep water.*
DISTRIBUTION *Mediterranean, Atlantic.*
SIMILAR SPECIES E. flavescens, *which is smaller, without a recessed front margin;* E. pennatifidum, *which is almost circular; and the tiny, flat* Echinocyamus pusillus, *which is no more than 1cm long.*

Maned Sea-slug

Aeolidia papillosa (Mollusca)

FOUND *at or below low tide zone on rocky shores, mudflats, and harbours.*

Many exclusively sublittoral sea-slugs are found around the coastlines of Europe, but the Maned Sea-slug is regularly found in the intertidal zone. Its name is derived from its dense covering of fleshy dorsal projections, or cerata. A predator of sea-anemones, it stores their stinging cells in the tips of its cerata for self-defence.

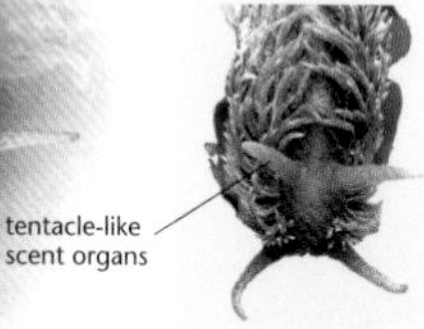

white-tipped, grey-brown projections

SIZE *Length to 12cm.*
FOOD *Eats sea-anemones, including the Beadlet Anemone (p.98).*
SHORE ZONE *Intertidal to shallow sublittoral.*
DISTRIBUTION *Atlantic, as far south as northern Spain.*
SIMILAR SPECIES Facelina coronata *has cerata in clusters;* Cuthona nana *too has cerata in clusters, but has much longer rear tentacles.*

Sea-lemon

Archidoris pseudoargus (Mollusca)

OCCURS *on the lower zone of rocky shores, below large boulders, and offshore into deep water.*

Quite similar to the duller terrestrial relatives of the more elaborate sea-slugs (above), the Sea-lemon is often yellow, but can be green, brown, white, or pink. There is also a bright red form known as var. *flammea*; all colour forms are mottled, perhaps for camouflage. The Sea-lemon is the most common intertidal sea-slug in Europe.

variable dark blotchy markings

warty, often yellowish skin

EGGS

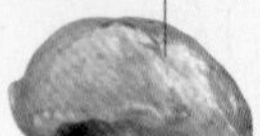

SIZE *Length to 12cm.*
FOOD *Eats sponges, especially Breadcrumb Sponge (p.93).*
SHORE ZONE *Middle intertidal to deep sublittoral.*
DISTRIBUTION *Mediterranean, Atlantic.*
SIMILAR SPECIES Geitodoris planata, *which is smaller, browner, with white star-shaped glands on its back.*

Laver Spire-shell

Hydrobia ulvae (Mollusca)

The mud-snails, the family to which the Laver Spire-shell belongs, are mostly all small and plain. They can be massively abundant: the Laver Spire-shell can be found in hundreds of thousands per square metre of estuarine mud. Collectively, this forms a major food resource for birds, especially surface pickers like Ringed Plover (p.190), and sifters like Shelduck (p.185).

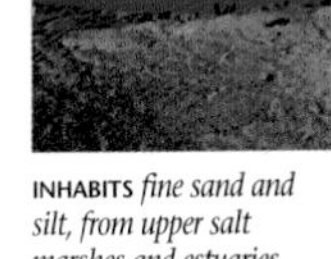

INHABITS *fine sand and silt, from upper salt marshes and estuaries to marine waters; tolerant of low and fluctuating salinity.*

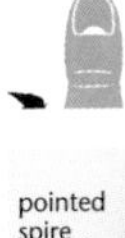

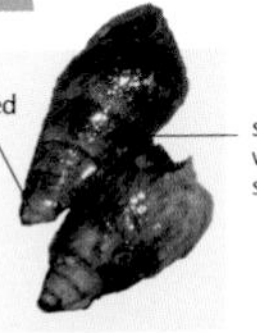

SIZE *Height to 6mm.*
FOOD *Grazes bacteria, diatoms (microscopic algae), and detritus.*
SHORE ZONE *Strandline to deep sublittoral.*
DISTRIBUTION *Throughout the region.*
SIMILAR SPECIES H. ventrosa, *which has more rounded whorls and a blunt apex;* H. neglecta, *which is smaller, and has a black "V" near the tentacle tips.*

Mediterranean Cone-shell

Conus ventricosus (Mollusca)

A distinctively shaped shell, with a short, sharp spire and a long, slit-like aperture, the Mediterranean Cone-shell is the only common member of its family in Europe. Cone-shells are renowned as predators, killing their prey by injecting them with a toxin. The shell usually has a network of olive, brown, or reddish flecks.

FAVOURS *rocky shores; often partially buried in the sand near rock faces.*

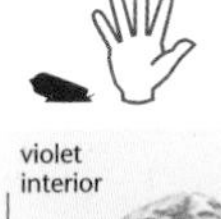

SIZE *Length to 7cm.*
FOOD *Predator of worms and other molluscs.*
SHORE ZONE *Intertidal to shallow sublittoral.*
DISTRIBUTION *Mediterranean, Atlantic (Portugal).*
SIMILAR SPECIES Columbella rustica, *which has a similar shape, but is only a third of the size; the outer lip projects out of the shell at the top of the aperture.*

Netted Dog-whelk

Hinia reticulata (Mollusca)

OCCURS at lower shore levels, in muddy patches between seaweed-covered rocks or in seagrass beds.

Sometimes known as *Nassarius reticulatus*, the Netted Dog-whelk has a characteristically sculptured shell, the net-like, criss-cross pattern caused by closely spaced ribs intersecting with spiral grooves. The shell is thick and shortly conical, with a spire of up to ten whorls. The outer lip of the shell is toothed. The Netted Dog-whelk lays vase-shaped egg capsules in rows on seaweeds, sea-grasses, and stones.

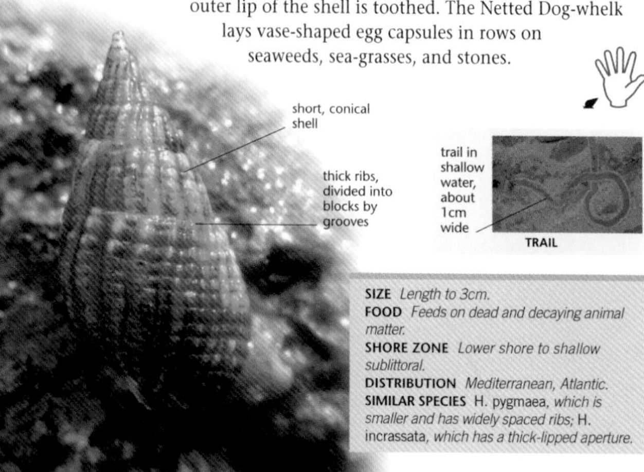

SIZE *Length to 3cm.*
FOOD *Feeds on dead and decaying animal matter.*
SHORE ZONE *Lower shore to shallow sublittoral.*
DISTRIBUTION *Mediterranean, Atlantic.*
SIMILAR SPECIES H. pygmaea, *which is smaller and has widely spaced ribs;* H. incrassata, *which has a thick-lipped aperture.*

Dog-whelk

Nucella lapillus (Mollusca)

FAVOURS all manner of rocky shores, except the most sheltered, seaweed-covered locations.

A thick-shelled, predatory snail, the Dog-whelk is usually whitish or pale yellow in colour, although it can be a darker brown, with dark bands on the shell. It is broadly conical with a short spire – the first (body) whorl occupies around three-quarters of its total length. It has an elongated, oval aperture, the margins of which are shallowly toothed. It produces distinctive, yellow egg capsules, which are fixed in clusters underneath rocks.

SIZE *Length to 3cm, sometimes longer.*
FOOD *Hunts barnacles, mussels, and other molluscs. Accesses larger prey items by boring a hole through the shell and inserting its proboscis through the hole.*
SHORE ZONE *Intertidal, from the neap high water to the low-water mark of spring tides.*
DISTRIBUTION *Atlantic.*
SIMILAR SPECIES *None.*

Common Whelk

Buccinum undatum (Mollusca)

A large, much-valued edible shellfish, the Common Whelk is fished commercially using traps baited with carrion. It has a tall, pointed shell, which is sculptured with wavy ribs crossed by spiral ridges and grooves. The egg masses are equally distinctive: round, spongy balls of lentil-shaped eggs, attached to subtidal rocks. Once empty, these are frequently washed ashore in the form of Sea Wash Balls (p.211).

INHABITS *a range of shorelines, from rocky to deposits of gravel, sand, or mud; also in estuaries.*

SIZE *Length to 11cm.*
FOOD *Scavenges on dead animals; preys on worms, cockles, and other bivalves.*
SHORE ZONE *Lower shore zone to offshore.*
DISTRIBUTION *Atlantic.*
SIMILAR SPECIES Neptunea antiqua, *which is larger, smooth and restricted to the sublittoral;* N. contraria, *which is from deep water, but is found on beaches as a fossil.*

Oyster Drill

Ocenebra erinacea (Mollusca)

Also known as European Sting-winkle, the Oyster Drill is a gastropod with a tall, angular spire and protruding ribs on the shell, which also bears deep, spiral striations. As its name suggests, it preys on oysters, among other species. It drills a hole in the shell of its prey, through which it injects a poison, killing it. The American Sting Winkle (*Urosalpinx cinerea*) is less angular, but a more serious pest of shellfish beds.

OCCURS *at the lower tidal levels of sheltered rocky shores, especially during the summer, but mainly sublittoral.*

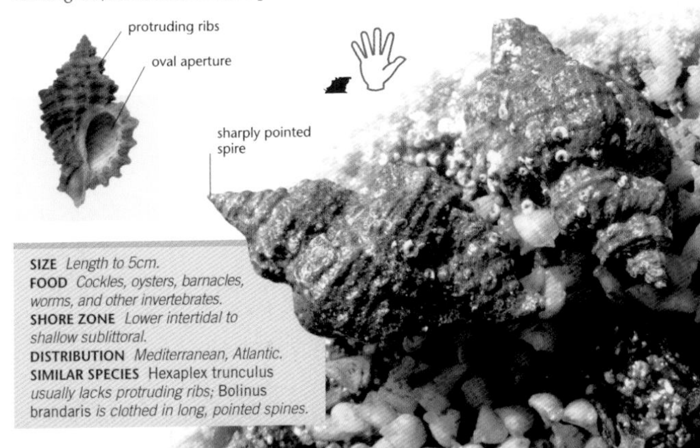

SIZE *Length to 5cm.*
FOOD *Cockles, oysters, barnacles, worms, and other invertebrates.*
SHORE ZONE *Lower intertidal to shallow sublittoral.*
DISTRIBUTION *Mediterranean, Atlantic.*
SIMILAR SPECIES Hexaplex trunculus *usually lacks protruding ribs;* Bolinus brandaris *is clothed in long, pointed spines.*

Common Periwinkle

Littorina littorea (Mollusca)

FAVOURS *rocky shores, especially with a good covering of seaweeds; also in areas with a sandy bottom, and in brackish estuaries.*

Periwinkles are the most abundant and some of the most familiar molluscs found on European rocky shores. All are largely intertidal – their thick, globular shells help them resist wave action, while the horny operculum ("door" that closes the shell) allows them to withstand long periods out of water. The largest, and often the commonest species, is the Common Periwinkle, which is dark grey-brown, finely striated, and has a sharply pointed shell. As individuals get older, the shell tends to become paler, and smoother as the striations become worn down.

SIZE *Height to 5cm, though usually smaller.*
FOOD *Grazes a wide range of seaweeds.*
SHORE ZONE *Upper intertidal down to the shallow sublittoral.*
DISTRIBUTION *Atlantic, north from northern Spain.*
SIMILAR SPECIES *Other periwinkle species are smaller and often coloured yellowish.* L. saxatilis *has more marked striations;* L. mariae *has its aperture broader than its body whorl;* L. obtusata *has an oval, flat-topped shell;* L. nigrolineata *has brown striations.* L. neritoides, *which is small and sharply pointed is the only species extending up to the Mediterranean.*

NOTE

Common Periwinkle is the most familiar edible winkle, and is collected commercially by hand. It requires careful boiling in order to kill pathogens, before the snail is safe for human consumption.

Grey Top-shell

Gibbula cineraria (Mollusca)

Distinctively coloured, with reddish bands on a grey or yellowish background, the Grey Top-shell often becomes rather worn around the apex, giving it a silvery appearance. The shell is pyramidal in shape and, as in many related species, has a distinct umbilicus, a small hole running from the centre of the base up towards the apex. In this species, the umbilicus has a marked oval shape; the size and shape of the umbilicus is an important distinguishing feature between species.

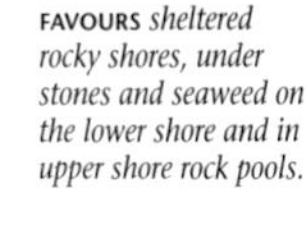

FAVOURS *sheltered rocky shores, under stones and seaweed on the lower shore and in upper shore rock pools.*

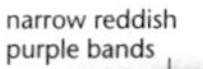

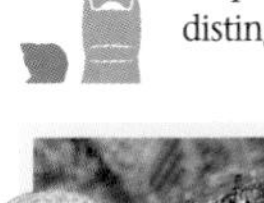

SPAWNING

SIZE *Height to 1.5cm.*
FOOD *Grazes algae.*
SHORE ZONE *Upper shore pools down to the deep sublittoral.*
DISTRIBUTION *Atlantic.*
SIMILAR SPECIES G. umbilicalis *is more blunt, with broader stripes;* G. pennanti *has broad stripes and a stepped profile;* Monodonta turbinata *has a toothed aperture.*

Painted Top-shell

Calliostoma zizyphinum (Mollusca)

Like many of the European Top-shells, the Painted Top-shell is often coloured with irregular red or brown bands spreading from the apex of the shell. It has a very characteristic shape, with a flat base and a perfectly conical spire. The shell has up to 12 whorls, each of which has numerous spiral striations. The living animal is brightly coloured – usually pinkish, flecked with red, brown, and yellow.

INHABITS *sheltered rocky shores with an extensive covering of seaweeds, and often found towards the low-water mark.*

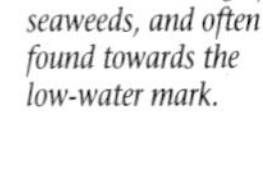

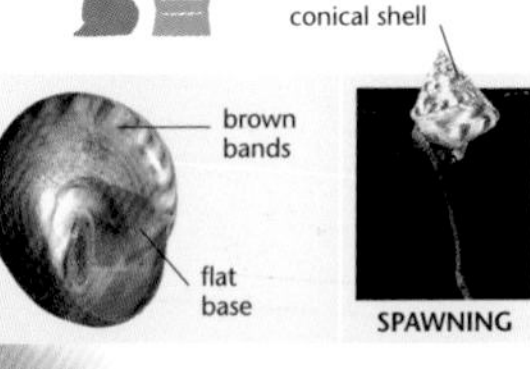

SPAWNING

SIZE *Length to 3cm.*
FOOD *Grazes algae and detritus.*
SHORE ZONE *Mid-shore zone to deep sublittoral.*
DISTRIBUTION *Mediterranean, Atlantic.*
SIMILAR SPECIES C. laugieri *(Mediterranean only), which is smaller with a steeper spire;* Monodonta *species have a more rounded profile and base, and a toothed aperture.*

Slipper-limpet

Crepidula fornicata (Mollusca)

ATTACHES *itself to stones and shells in hard but muddy estuarine and open coastal habitats.*

Introduced with commercial oysters from North America around 1890, the Slipper-limpet is now well established in Europe. In some places, it is a serious pest of shell fisheries as it carpets large areas of lower tidal mud. Slipper-limpets often stick together to form arched chains; they begin life as males, but as they grow and others add to the chain above them, they change sex progressively to become female, a process which takes about 60 days to complete.

rounded, humped profile

smooth, pinkish brown shell

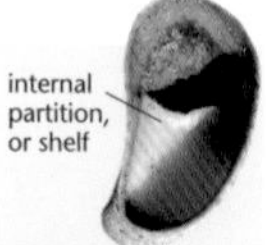

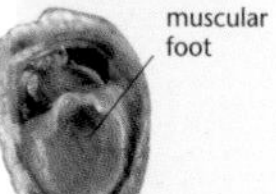

SIZE *Length to 5cm; height to 2.5cm.*
FOOD *Filters micro-organisms and organic detritus from sea water; binds food with mucus.*
SHORE ZONE *Mean low-water mark to the shallow sublittoral.*
DISTRIBUTION *Mediterranean (local); Atlantic, Spain to Denmark.*
SIMILAR SPECIES C. gibbosa, *which has a thin shell, with a shorter internal shelf.*

Green Ormer

Haliotis tuberculata (Mollusca)

INHABITS *lower levels of rocky shores, especially rocks covered with encrusting red seaweeds.*

A large, fleshy mollusc, the Green Ormer clings tightly to rocks with its powerful muscular foot, but is capable of a remarkable turn of speed when threatened. Highly prized as food, overfishing has led to its severe decline in places. Typically, these molluscs are collected by hand at the lowest tides, known as "ormering tides". Under water, portions of the white living mantle may protrude through holes in the shell.

line of raised, round holes

mother-of-pearl lining

flattened, ear-shaped shell

coiled apex

brown, with reddish or green tinge

SIZE *Length to 10cm; height to 2cm.*
FOOD *Grazes small red and green seaweeds.*
SHORE ZONE *Mean low-water mark to the shallow sublittoral.*
DISTRIBUTION *Mediterranean; Atlantic, north to the English Channel.*
SIMILAR SPECIES H. lamellosa, *which is smaller, flatter, and often encrusted with algae and other organisms; Mediterranean only.*

Chinaman's Hat

Calyptraea chinensis (Mollusca)

Although similar in appearance to a small limpet (p.120), Chinaman's Hat has a shelf or partition across part of its shell aperture, indicating that it belongs to the same family as the Slipper-limpet (p.118). Unlike its relatives, however, the edge of the shelf is strongly curved. The shell exterior is always pale whitish or yellow, while the inner surface is pearly white. This species is believed to be spreading northwards due to rising sea temperatures as a result of global climate change.

FOUND *on shells and under stones at the lower tidal levels of relatively sheltered coasts.*

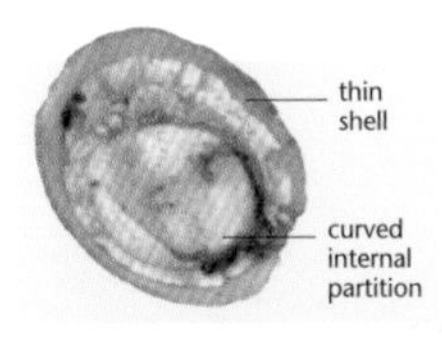

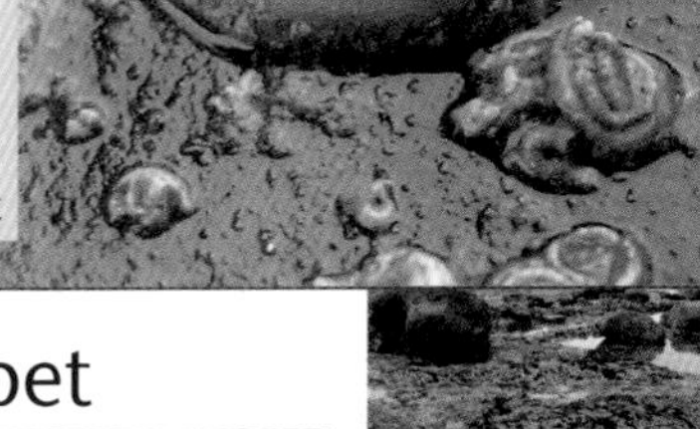

SIZE *Height to 5mm; width to 1.5cm.*
FOOD *Filter-feeding on organic matter.*
SHORE ZONE *Mean low-water mark to the shallow sublittoral.*
DISTRIBUTION *Mediterranean; Atlantic, north to western Britain.*
SIMILAR SPECIES *Unrelated limpets and Tortoiseshell-limpets (p.120) have a similar shape, but are thicker, without an internal shelf.*

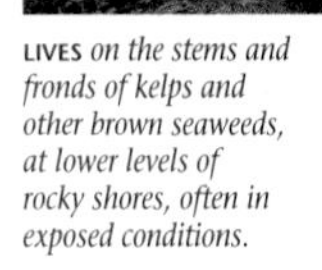

Blue-rayed Limpet

Helcion pellucidum (Mollusca)

With its radiating lines of pale blue spots on a translucent amber shell, the Blue-rayed Limpet is a beautiful mollusc. It feeds on brown seaweeds and creates its own place of shelter by excavating cavities in the stalks of the larger kelp species. Such refuge enables it to thrive in exposed conditions, despite its thin, delicate shell. A short-lived species, it rarely lives for more than a year.

LIVES *on the stems and fronds of kelps and other brown seaweeds, at lower levels of rocky shores, often in exposed conditions.*

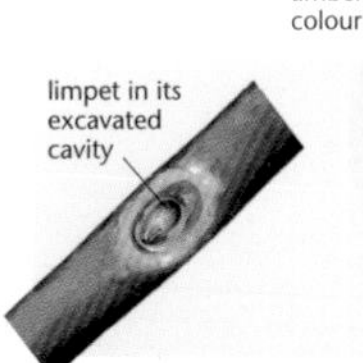

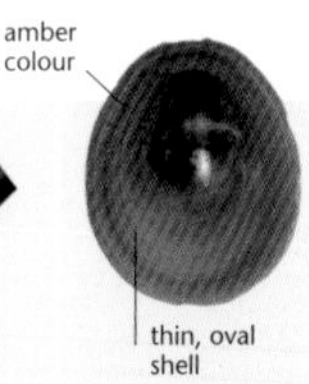

SIZE *Length to 1.5cm.*
FOOD *Grazes brown seaweeds.*
SHORE ZONE *Lower intertidal zone to the sublittoral.*
DISTRIBUTION *Atlantic.*
SIMILAR SPECIES *Chinaman's Hat (above), which has a paler shell, and a more sharply pointed and centrally placed apex, as well as an internal shell partition.*

Common Limpet

Patella vulgata (Mollusca)

STICKS *securely on intertidal rocks, especially on exposed rocky shores; not common where there is abundant seaweed.*

Supremely adapted to intertidal life, the thick shell of the Common Limpet and its ability to cling tightly to hard surfaces help it to withstand wave attack, exposure to the air, and predation from birds and fish alike. It forages underwater, following a "snail trail" of mucus back to its home base. The Common Limpet is long-lived (up to 15 years), and forms a small depression, known as a scar, by rubbing against the rock it attaches itself to.

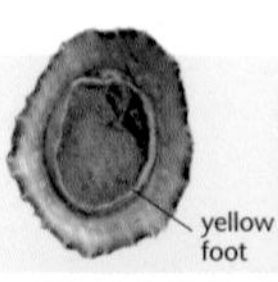

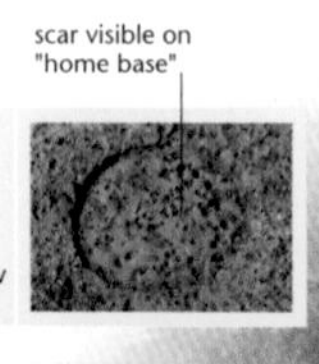

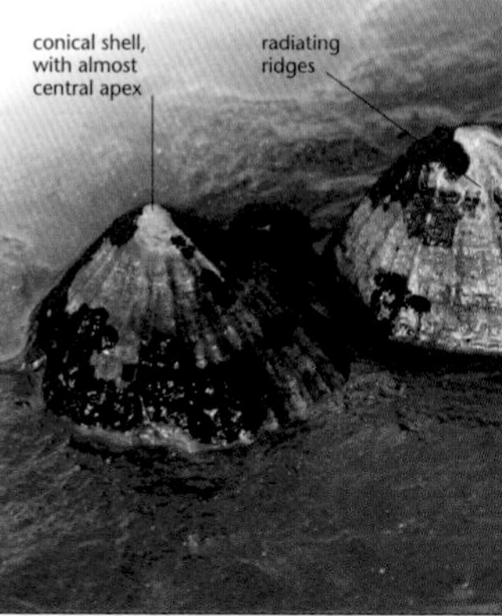

SIZE *Length to 6cm.*
FOOD *Grazes algae, especially newly germinated sporelings.*
SHORE ZONE *Intertidal.*
DSTRIBUTION *Mediterranean, Atlantic.*
SIMILAR SPECIES P. depressa, *which is smaller, flatter, and darker, and has a dark brown foot;* P. caerulea *(Mediterranean), which has deep ribs and an irregular shell edge.*

White Tortoiseshell-limpet

Tectura virginea (Mollusca)

INHABITS *rocky shores at lower intertidal levels, especially among encrusting red algae.*

Tortoiseshell-limpets differ from the true limpets in their small size, low conical profile, and asymmetric peak. They feed primarily on encrusting red seaweeds. The White Tortoiseshell-limpet is the smaller and paler of the two European species, the other being *T. testudinalis*. It has an unmarked white or pale pink shell, sometimes with darker rays towards the shell margin. Although they are relatively short-lived, their shells may become a substrate for the algae that they feed on.

apex just off-centre

low conical profile

SIZE *Length to 1.5cm; height to 6mm.*
FOOD *Grazes encrusting red algae, and possibly sponges.*
SHORE ZONE *Mean low water to sublittoral.*
DISTRIBUTION *Mediterranean, Atlantic.*
SIMILAR SPECIES T. testudinalis, *which is larger, to 3cm long, and usually with dark mottling. It is a northern Atlantic species, extending as far south as northern Britain.*

Common Keyhole-limpet

Diodora graeca (Mollusca)

This keyhole-limpet has an oval shell with a small hole at the slightly off-centre apex, through which the limpet exhales a stream of water. Its surface has a sculpted look due to the intersection of very prominent ribs radiating from the apex and numerous concentric striations; the ribs often protrude from the shell edge, producing a frilly margin.

FAVOURS *the lower levels of rocky shores, where its main food species can be found.*

SIZE *Length to 4cm; height to 1cm.*
FOOD *Eats sponges such as Breadcrumb Sponge (p.93).*
SHORE ZONE *Low intertidal to deep offshore.*
DISTRIBUTION *Mediterranean; Atlantic, north to Britain.*
SIMILAR SPECIES D. gibberula *has a very asymmetric profile;* D. italica *has finer ribs;* Puncturella noachina *is more steeply conical.*

Common Slit-limpet

Emarginula fissura (Mollusca)

Belonging to the same family as the keyhole-limpets (above), slit-limpets have their exhalant aperture at the shell margin. The Common Slit-limpet is a small, but tall-spired species, with its apex curved backwards; its slit continues as a groove right to the apex. The shell is generally white or pale yellowish. It has a prominent chequered surface formed by the radiating ribs intersecting with spiral striations.

FOUND *on rocks and under boulders at the lower levels of a rocky shore, extending well into the sublittoral.*

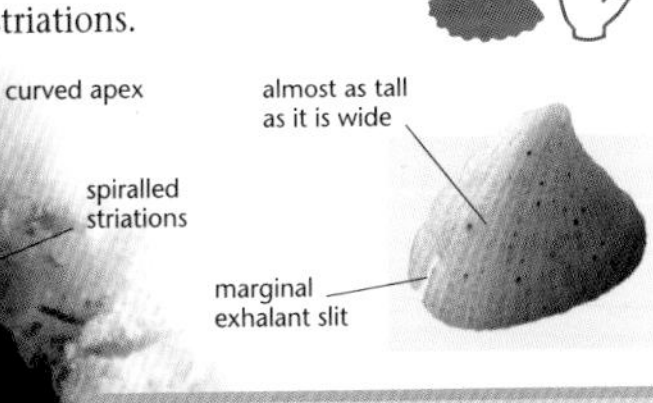

SIZE *Length to 1cm; height to 8mm.*
FOOD *Eats sponges.*
SHORE ZONE *Lower intertidal to sublittoral.*
DISTRIBUTION *Mediterranean, Atlantic.*
SIMILAR SPECIES E. crassa, *which is larger, with indistinct ridges;* E. conica, *which is smaller, with a very recurved apex; several other species are restricted to the Mediterranean.*

Spotted Cowrie

Trivia monacha (Mollusca)

OCCURS *at lower tidal levels on rocky shores, usually around colonies of its prey species.*

The cowries are a group of distinctive, egg-shaped molluscs, with a narrow, elongated aperture running almost the whole length of the shell. Largely found in subtropical waters, there are just five European species, the most common of which is the Spotted Cowrie, distinguished by the three diffuse, dark spots on the shell. The female Spotted Cowrie deposits egg capsules in holes eaten out of a sea-squirt.

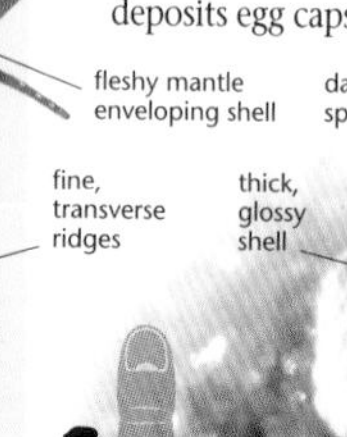

SIZE *Length to 1.2cm; width to 8mm.*
FOOD *Small colonial sea-squirts.*
SHORE ZONE *Lower intertidal to the shallow sublittoral.*
DISTRIBUTION *Mediterranean; Atlantic, north to Britain.*
SIMILAR SPECIES T. arctica, *which is smaller and unspotted;* T. multilirata, *which is larger and more globular.*

Callochiton septemvalvis

Callochiton septemvalvis (Mollusca)

FAVOURS *the lower zones of rocky shores, usually on rocks covered with seaweed.*

All chitons, or coat-of-mail shells, have a similar appearance – the shell is made up of eight transverse plates within a broad girdle. They clamp themselves tightly to rocks, using their muscular foot. The various species are distinguished by the raised pattern on their fringing girdle and by the shape of the plate edges. These, however, can only be seen after the chiton dies, as the plate edges overlap each other.

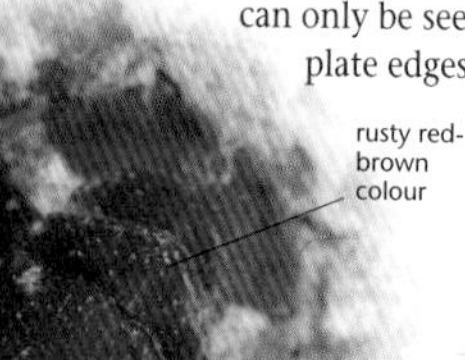

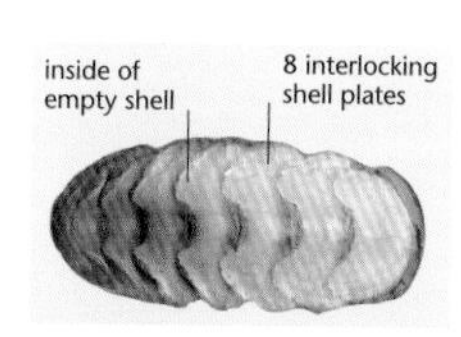

SIZE *Length to 3cm.*
FOOD *Grazes small and encrusting seaweeds, and bryozoa.*
SHORE ZONE *Lower intertidal zone to shallow sublittoral.*
DISTRIBUTION *Mediterranean, Atlantic.*
SIMILAR SPECIES *All 40 species of chiton in European coastal waters have a similar body and shell form; many only sublittoral.*

Common Mussel

Mytilus edulis (Mollusca)

A long-lived bivalve mollusc of commercial importance, the Common Mussel is valued as food by humans, fish, starfish, crabs, and birds alike. Its shell is sculpted with concentric lines, the interior being pearly white with a purplish border. The umbo on the shell is off-centre, pointed, and almost straight. Sea ducks, such as Eiders (p.182) dive for mussels when the tide is in, using their strong neck muscles to wrench them off the rocks.

ATTACHES *to rocks and other hard substrates on rocky coasts, often forming dense beds.*

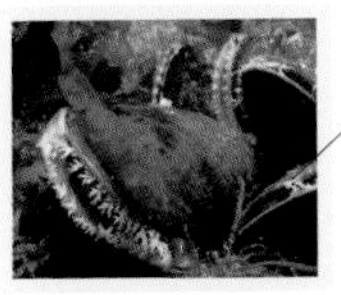

SIZE *Length to 10cm, sometimes longer.*
FOOD *Filters micro-organisms and detritus.*
SHORE ZONE *Middle and lower intertidal to shallow sublittoral.*
DISTRIBUTION *Atlantic, north from northern Spain.*
SIMILAR SPECIES M. galloprovincialis *has a downcurved umbo;* Modiolus modiolus *is larger, with a broad shell and blunt umbo.*

Fan-mussel

Pinna nobilis (Mollusca)

A very large bivalve, recorded up to a metre long, the Fan-mussel has a shell that looks like a folded fan, although its width varies considerably. It has suffered a major decline in numbers as a result of over-collection and habitat damage, and is now strictly protected. Larger, smoother specimens are sometimes referred to as *P. squamosa*, but they are probably just older example of this species.

LIVES *upright in sand and gravelly shores, its broad end protruding; often in seagrass beds.*

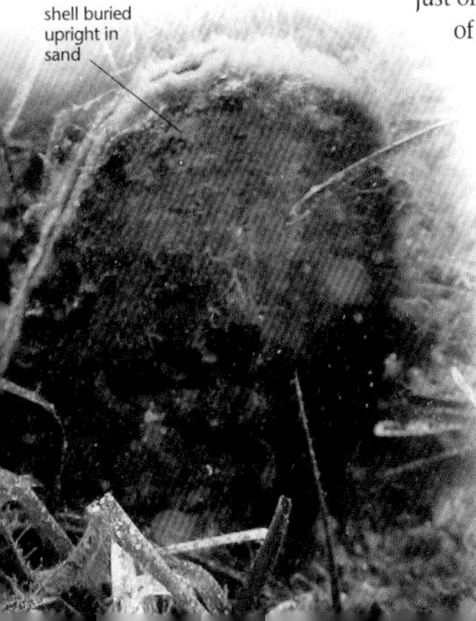

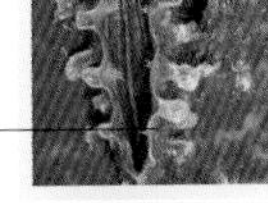
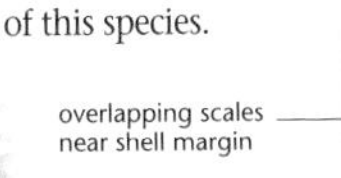

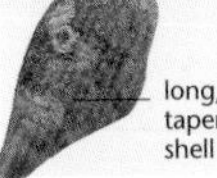

SIZE *Length to 45cm, sometimes longer.*
FOOD *Filters organic matter from sea water.*
SHORE ZONE *Lower intertidal to shallow sublittoral.*
DISTRIBUTION *Mediterranean.*
SIMILAR SPECIES Atrina fragilis, *which is smaller, usually paler, and has radiating ribs;* Lima *species, which have a similar fan shape, but a truncated umbo, and are much smaller.*

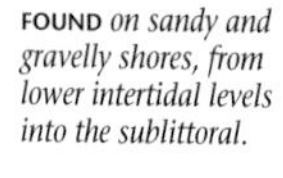

Queen Scallop

Aequipecten opercularis (Mollusca)

FOUND *on sandy and gravelly shores, from lower intertidal levels into the sublittoral.*

A free-living species as an adult, although attached to a stone by a byssus as a juvenile, the Queen Scallop slowly flaps its shell to move forward, and uses jet propulsion as an escape mechanism. The fan-shaped shell has about 20 prominent, radiating ribs, which extend at the margin as shallow teeth, and slightly unequal ears at the base.

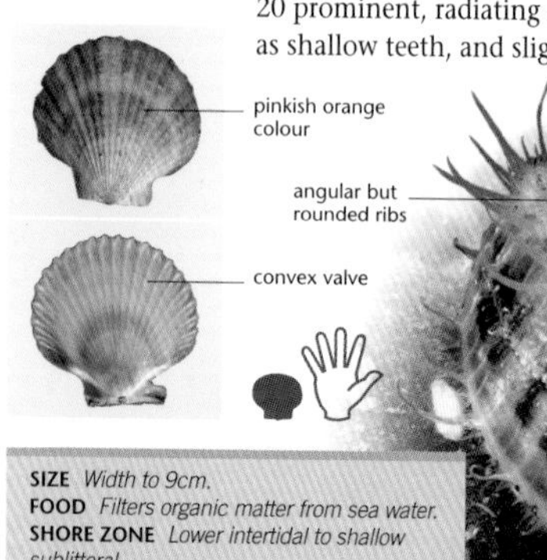

SIZE *Width to 9cm.*
FOOD *Filters organic matter from sea water.*
SHORE ZONE *Lower intertidal to shallow sublittoral.*
DISTRIBUTION *Mediterranean, Atlantic.*
SIMILAR SPECIES Pecten maximus, *which is larger, with fewer rounded ribs;* P. jacobaeus, *which has very broad, square-sectioned ribs, wider than the adjacent grooves.*

Variegated Scallop

Chlamys varia (Mollusca)

INHABITS *rocky shores, free-living or attached to rocks or seaweed holdfasts by a byssus.*

A rather elongated scallop of rocky shorelines, the Variegated, or Black, Scallop is a sequential hermaphrodite, maturing as a male then changing sex several times during its life. The shell is very variable in colour, ranging from purple and orange to yellow and white, with all shades in between, often in irregular patterns. The ribs on the shell bear spines, both of which are especially well-developed near the margin.

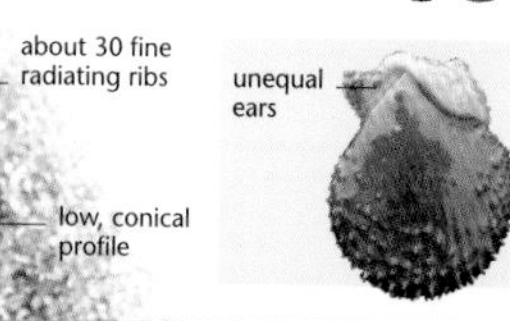

SIZE *Length to 6cm.*
FOOD *Filters organic matter from sea water.*
SHORE ZONE *Lower intertidal to deep sublittoral.*
DISTRIBUTION *Mediterranean, Atlantic.*
SIMILAR SPECIES C. distorta, *which has finer ribs, and the lower valve cemented to a rock;* Palliolum tigerinum, *which is only to 2.5cm wide, and has smooth shell valves.*

Mediterranean Jewel-box

Chama gryphoides (Mollusca)

A member of a largely tropical family, renowned for its often bizarrely sculpted shell surface and much prized by collectors, the Mediterranean Jewel-box is adorned by a series of coarse, concentric rings, sometimes raised into projecting flakes. The deeply cupped lower shell is cemented to a rock surface, while the upper shell is fairly flat. The umbo shows a slight but distinct tendency to spiral.

LIVES *cemented to rocks and stones on Mediterranean rocky shores, in the lower intertidal zone.*

SIZE *Length to 4cm.*
FOOD *Filters organic matter from sea water.*
SHORE ZONE *Lower intertidal to shallow sublittoral.*
DISTRIBUTION *Mediterranean.*
SIMILAR SPECIES *The unrelated Common Oyster and* Crassostrea gigas *(p.126) are similarly sculpted, but are usually larger, with a single large muscle scar inside each shell.*

Noah's Ark-shell

Arca noae (Mollusca)

ATTACHES *by a byssus to rocks, often in crevices, in the intertidal zone of rocky shores.*

Ark-shells are a family of angular, thick-shelled molluscs, in which the hinge-line extends for almost the full length of the shell. Noah's Ark-shell is distinctive in shape and appearance, with a velvet-hairy outer layer to the shell (the periostracum). It is often encrusted by a sponge *Crambe crambe*, which protects it from predation.

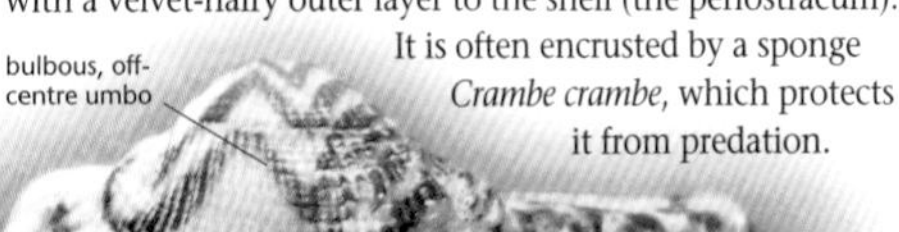

SIZE *Length to 8cm.*
FOOD *Filters organic matter from sea water.*
SHORE ZONE *Middle intertidal to shallow sublittoral.*
DISTRIBUTION *Mediterranean, Atlantic.*
SIMILAR SPECIES A. tetragona, *which is smaller, more angular, and more widespread;* Striarca lactea, *which is smaller still and finely sculpted, with a central umbo.*

Common Oyster

Ostrea edulis (Mollusca)

FORMS *beds around the low-water mark of estuaries and open shores, often attached to small rocks or other shells.*

Also called Flat Oyster or Native Oyster (it is the only species native to Britain and northwest Europe), the Common Oyster is renowned as a source of food and pearls. Its population has fallen sharply due to overfishing, pollution, disease, and competition from Slipper-limpets (p.118). It is a bivalve, the shells differing in size and shape. The larger, curved, left valve is attached to and fused with the substrate.

SIZE *Diameter to 10cm.*
FOOD *Filters organic matter from sea water.*
SHORE ZONE *Extreme low-water mark to sublittoral.*
DISTRIBUTION *Mediterranean, Atlantic.*
SIMILAR SPECIES Crassostrea gigas, *which is larger, often longer and narrower; it is a native species as far north as the bay of Biscay and introduced further north.*

Saddle-oyster

Anomia ephippium (Mollusca)

ATTACHES *itself to rocks, algal holdfasts, and shells, in the lower zone of rocky shores.*

Despite a similar appearance, the Saddle-oyster is not closely related to the true oysters. One significant difference is that the domed left valve is uppermost *in situ*, and the thin, flat right valve has a hole near the hinge, through which the attachment structure (byssus) passes to the substrate. It has a thin shell with concentric ridges and scales. The inner surface of the upper shell has three closely spaced muscle scars.

SIZE *Diameter to 6cm.*
FOOD *Filters organic matter from sea water.*
SHORE ZONE *Lower intertidal to shallow sublittoral.*
DISTRIBUTION *Mediterranean, Atlantic.*
SIMILAR SPECIES *Other saddle-oysters, such as* Monia squama, *which are usually smaller and more regularly rounded, some being exclusively sublittoral.*

Common Cockle

Cerastoderma edule (Mollusca)

A favoured food of several wading birds, and also widely fished for human consumption, the Common Cockle forms dense populations just below the surface of sand or mud. It lives for four or five years, its ridged shells showing distinct concentric annual growth rings. The Common Cockle is tolerant of reduced salinity in estuaries, to one third the salt concentration found in sea water.

BURROWS *in the upper 5cm of sandy, muddy, and fine gravel shores.*

upper and lower shells almost identical

rounded and humped shell

about 24 radiating ribs

fleshy foot

SIZE *To 5cm long.*
FOOD *Filters organic matter from sea water.*
SHORE ZONE *Middle intertidal down to just below the low-water mark.*
DISTRIBUTION *Mediterranean, Atlantic.*
SIMILAR SPECIES *Rough Cockle (below);* C. glaucum, *which prefers low salinity areas and is found in the Baltic and brackish lagoon habitats.*

Rough Cockle

Acanthocardia tuberculata (Mollusca)

A large cockle with a southerly distribution, the Rough Cockle is one of several species bearing spines or tubercles on the ribs. The spiny species are differentiated on the basis of the size, shape, and distribution of the spines, and the structures within the hinge of the shell. The shell of the Rough Cockle is banded dark brown or is brown all over; the inside of the shell is white, with grooves running down one-third of the way from the margin to the hinge.

LIVES *in muddy sand and gravel, burrowing just below the surface, on the lower shore of open coasts.*

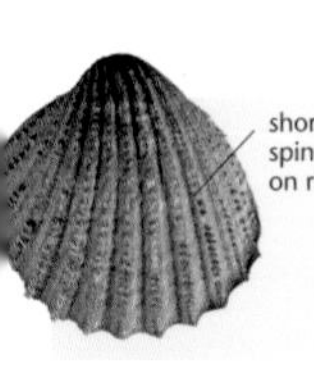

short spines on ribs

white interior of shell

18–20 keeled ribs

SIZE *To 9cm long.*
FOOD *Filters organic matter from sea water.*
SHORE ZONE *Lower intertidal to shallow sublittoral.*
DISTRIBUTION *Mediterranean; Atlantic, north to southern Britain.*
SIMILAR SPECIES A. aculeata *and* A. echinata, *which have longer, sharper spines along the ribs of the shell.*

Striped Venus-clam

Chamelea gallina (Mollusca)

BURIES *itself in fine to coarse sand, at lower levels of sand-flats on the open coast and the outer reaches of estuaries.*

An important source of food, especially in Mediterranean countries, the Striped Venus-clam is dredged from the sandy sediments in which it is buried, from the intertidal to the fairly deep sublittoral zone. The background colour is pale cream or yellowish, usually marked by three prominent, brown bands that radiate out from the off-centre umbo, and fine, concentric ridges.

thick, triangular-oval shell

broad, chestnut-coloured rays

SIZE *Length to 4cm.*
FOOD *Filters organic matter from sea water.*
SHORE ZONE *Lower shore to sublittoral.*
DISTRIBUTION *Mediterranean, Atlantic.*
SIMILAR SPECIES Venus verrucosa, *which has coarse ridges cut by radiating grooves;* Callista chione, *which is larger, with more, but narrower coloured rays;* Dosinia *species, which have a hooked umbo, but on a circular shell.*

Banded Carpet-shell

Tapes rhomboides (Mollusca)

FOUND *shallowly buried in sand, from intertidal sandflats to deep water at the edge of the continental shelf.*

Also known by a variety of other generic names, including *Paphia* and *Venerupis*, the Banded Carpet-shell is often an abundant species in coarse sand and gravel intertidal habitats. It is very similar to the related Striped Venus-clam (above), with its off-centre umbo, numerous concentric grooves, and reddish brown rays of colour, but in general form, it is much more elongated.

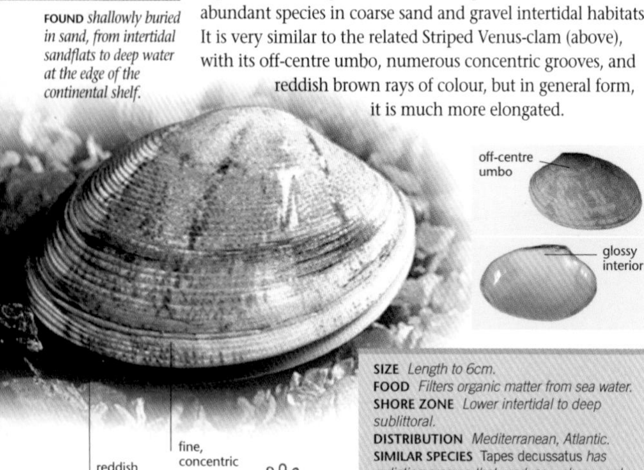

SIZE *Length to 6cm.*
FOOD *Filters organic matter from sea water.*
SHORE ZONE *Lower intertidal to deep sublittoral.*
DISTRIBUTION *Mediterranean, Atlantic.*
SIMILAR SPECIES Tapes decussatus *has radiating grooves that produce a chequered effect; in* Venerupis senegalensis, *the radiating grooves are prominent at narrow end of shell.*

Thin Tellin

Angulus tenuis (Mollusca)

Often very abundant in its chosen habitat, the Thin Tellin is an important source of food for wading birds and fish – young Plaice (p.160), for example, graze the siphons that it sticks out for feeding. Its shells are whitish, or a delicate pink or yellow, both on the inside and the outside. Despite their brittle texture, unbroken dead shells are often washed up on the strandline. The pink shells in particular are sometimes referred to as "Babies' Toenails".

OCCUPIES *extensive flats of fine to medium sand, shallowly buried in the surface layer, to a depth of about 10cm.*

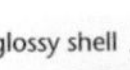

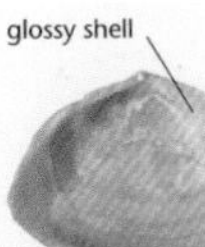

SIZE *Length to 2.5cm.*
FOOD *Ingests deposited organic matter through its long siphon.*
SHORE ZONE *Middle intertidal to shallow sublittoral.*
DISTRIBUTION *Mediterranean, Atlantic.*
SIMILAR SPECIES A. squalidus *has a concave angle to its shell margin;* Fabulina fabula *has one end of the shell abruptly tapered.*

Baltic Tellin

Macoma balthica (Mollusca)

Like others in its family, the Baltic Tellin has an unusual feeding style for a bivalve. It uses its long siphon to hoover up deposited material from the surface of the mud, rather than filter organic matter from the water column. It has a plump, almost circular shell, which is variable in colour, from white to purple, although the true colour is often masked by the blackish mud in which it lies shallowly buried. The edge of the shell often has a papery, pale brown outer coat, or periostracum.

ABOUNDS *in estuaries and tidal mudflats, buried in mud and muddy sand.*

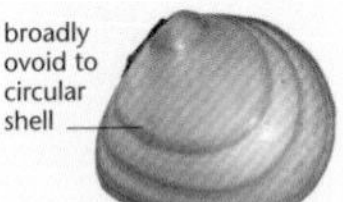

SIZE *Length to 2.5cm.*
FOOD *Ingests deposited organic matter through its long siphon.*
SHORE ZONE *Upper intertidal to shallow sublittoral.*
DISTRIBUTION *Atlantic, Baltic.*
SIMILAR SPECIES *Peppery Furrow-shell (p.130), which is larger and white inside;* Arcopagia crassa, *which has radiating red rays.*

Peppery Furrow-shell

Scrobicularia plana (Mollusca)

FAVOURS *sheltered, brackish habitats, making burrows to a depth of 20cm in mud and fine sand.*

Like the tellins (p.129), the Peppery Furrow-shell is a deposit-feeder, and its feeding activity is apparent at low tide from the star-shaped patterns made by the siphon around the burrow. When the inhalent siphon is extended, it is taken as food by crabs, fish, and wading birds, but the damaged tissue is replaced quickly, in around 5 days. Whole Peppery Furrow-shells are also preyed upon by some waders.

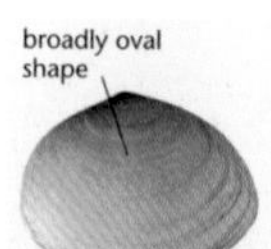

SIZE *Length to 6.5cm.*
FOOD *Feeds on organic deposits, using an extended siphon.*
SHORE ZONE *Intertidal.*
DISTRIBUTION *Mediterranean, Atlantic, Baltic.*
SIMILAR SPECIES *Baltic Tellin (p.129), which is similar but much smaller;* Arcopagia crassa, *which has radiating red rays on the shell.*

Wedge-shell

Donax trunculus (Mollusca)

INHABITS *moderately exposed sandy shores, living in shallow burrows.*

A bivalve of exposed sandy beaches, the Wedge-shell is sometimes known as the Surf-clam, from its habit of burrowing rapidly into the sand after being brought to the surface by rough surf. It is preyed upon by both wading birds and carnivorous gastropods, hence the number of dead specimens with perforated shells that can be found. The wedge-shaped shell usually has a purple interior.

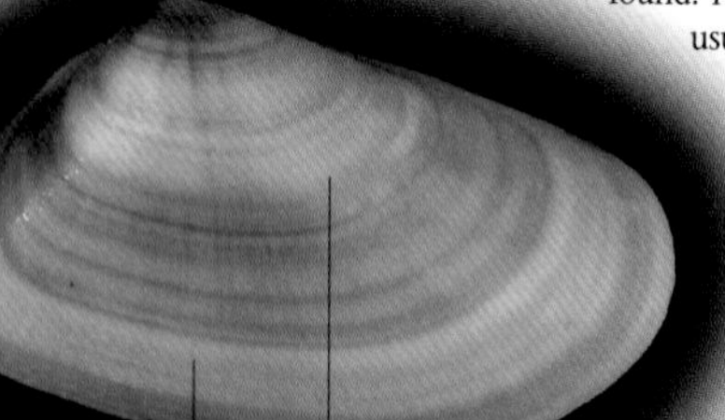

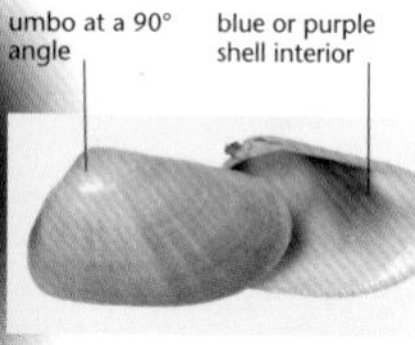

SIZE *Length to 4cm.*
FOOD *Filters organic matter from sea water.*
SHORE ZONE *Intertidal to shallow sublittoral.*
DISTRIBUTION *Mediterranean; Atlantic, north to Brittany.*
SIMILAR SPECIES D. vittatus, *which often has a yellow or orange shell interior, and pale growth rings;* D. variegatus, *which has a single, pale, radiating band of colour.*

Common Otter-shell

Lutraria lutraria (Mollusca)

A large, deep-burrowing bivalve, the Common Otter-shell maintains contact with the surface by means of a pair of large, fleshy, fused siphons, which cannot be fully retracted into the shell due to their size. The surface of the shell is usually obscured by extensive remnants of the periostracum (papery outer covering), which is dark brown in colour. This outer coat may be stained black by the mud that it lives in.

INHABITS *deep, muddy sandflats on the lower shore, burrowing to a depth of about 40cm.*

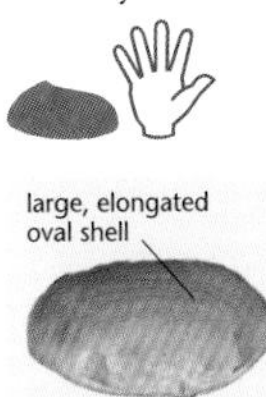

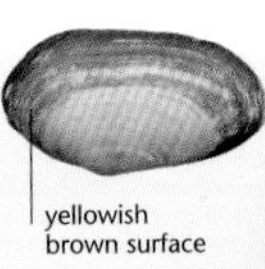

SIZE *Length to 13cm.*
FOOD *Filters organic matter from sea water.*
SHORE ZONE *Lower intertidal to sublittoral.*
DISTRIBUTION *Mediterranean, Atlantic.*
SIMILAR SPECIES L. oblonga, *which has a concave hinge-line and a highly off-centre umbo, and is found in the Mediterranean;* L. angustior, *which is more elongated and angular.*

Sand-gaper

Mya arenaria (Mollusca)

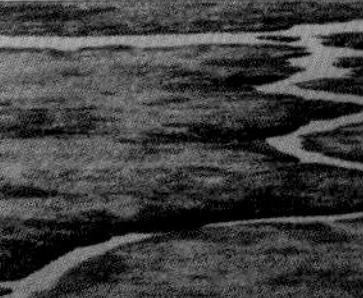

A cool-water species, the Sand-gaper is a large, coarsely sculpted bivalve, which burrows deeply into estuarine sediments, to a depth of 50cm. It uses its long, thick, fused siphons to draw in water from which it filter-feeds on organic matter, and leaves behind a distinctive keyhole-shaped opening on the surface of the sediment.

BURROWS *deeply into mud- and sandflats, especially in sheltered estuarine areas; tolerant of low salinity.*

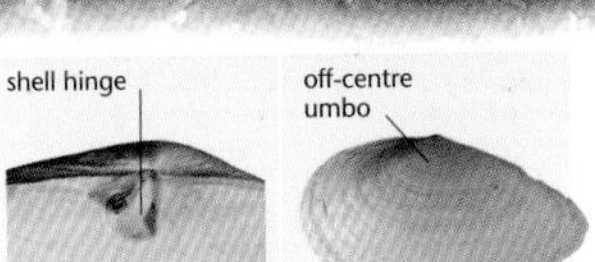

SIZE *Length to 15cm.*
FOOD *Filters organic matter from sea water.*
SHORE ZONE *Lower intertidal to shallow sublittoral.*
DISTRIBUTION *Atlantic, Baltic.*
SIMILAR SPECIES M. truncata *is smaller, with one sharply truncated end, hence its English name of Blunt Gaper;* Sphenia binghami *is also truncated, but only up to 2cm long.*

White Piddock

Barnea candida (Mollusca)

BORES *its shell into wood, peat, clay, and other soft rocks at lower tidal levels and in the sublittoral.*

Also known as Angel's Wings, from the colour and shape of the two valves of a dead specimen, the White Piddock is a borer, using its sharply sculpted shell to hollow out a hole in soft rock or wood. The tips of the shell often protrude from the hole, and show a pronounced gape, as they do not close completely. The foot muscle is too large to retract wholly within the shell. The filter-feeding siphons are equally large, up to five times the length of the shell.

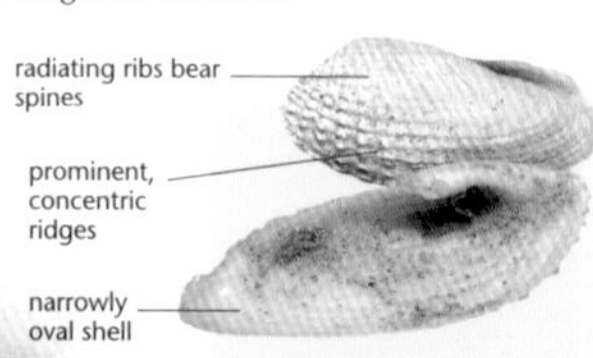

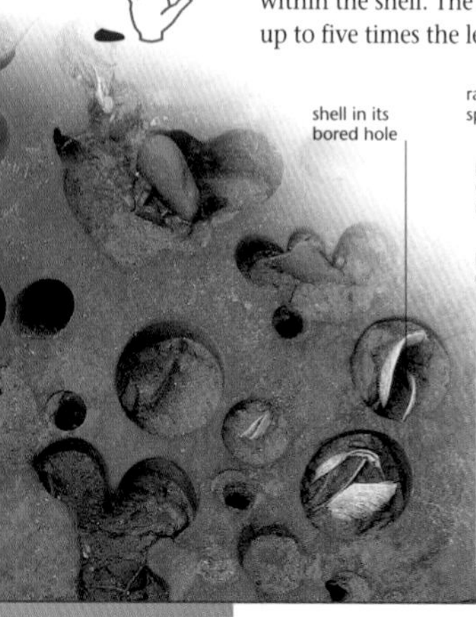

SIZE *Length to 6.5cm.*
FOOD *Filters organic matter from sea water.*
SHORE ZONE *Lower intertidal to shallow sublittoral.*
DISTRIBUTION *Mediterranean, Atlantic.*
SIMILAR SPECIES Pholas dactylus, *which has a concave margin to its shell, close to the hinge;* Petricola pholadiformis, *which is browner, with more prominent radiating ribs.*

Common Razor-shell

Ensis ensis (Mollusca)

LIVES *in permanent burrows in firm sand at lower shore levels; may be very abundant in sheltered locations.*

The razor-shells are elongated molluscs that live in vertical burrows in sand. When covered by the tide, their siphons protrude for feeding and respiration, but if threatened, the muscular foot rapidly pulls the whole shell deep into the burrow, leaving a keyhole-shaped entrance hole in the sand. The Common Razor-shell is distinguished by its slender, smoothly curved shell and its reddish foot.

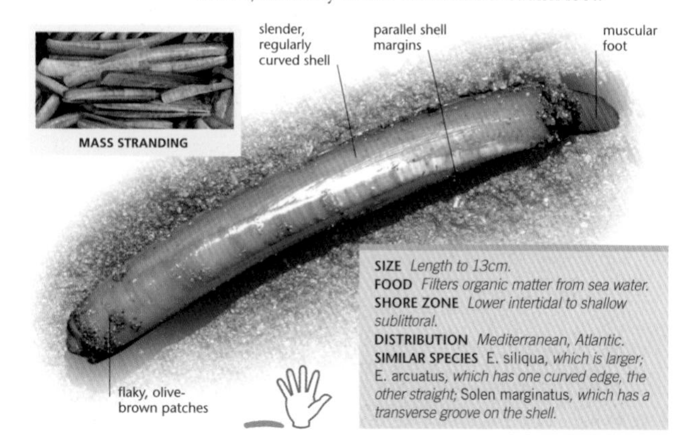

MASS STRANDING

SIZE *Length to 13cm.*
FOOD *Filters organic matter from sea water.*
SHORE ZONE *Lower intertidal to shallow sublittoral.*
DISTRIBUTION *Mediterranean, Atlantic.*
SIMILAR SPECIES E. siliqua, *which is larger;* E. arcuatus, *which has one curved edge, the other straight;* Solen marginatus, *which has a transverse groove on the shell.*

Northern Acorn-barnacle

Semibalanus balanoides (Crustacea)

Notwithstanding their apparent similarity to small molluscs, barnacles are actually crustaceans – a fact best appreciated by looking at the planktonic larvae, which resemble those of crabs. The adults are enveloped in a chalky shell made up of a number of fused plates, which form an aperture. This aperture is closed by the plates when out of the water. The Northern Acorn-barnacle is the most widespread species in the north of the region, and has a membranous shell-base, unlike other species that have calcified bases. The shape of the aperture and details of the plates are important identification markers for the species.

LIVES *permanently attached to rocks in the intertidal zone as an adult; can colonize estuarine waters with lower salinity.*

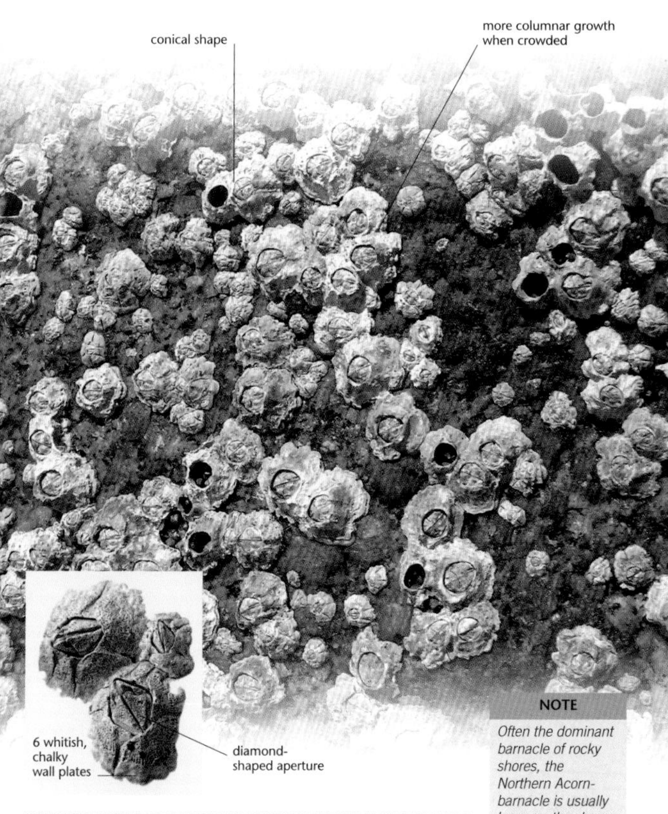

NOTE

Often the dominant barnacle of rocky shores, the Northern Acorn-barnacle is usually lower on the shore than the related Chthamalus montagui. *It is a northern species, whereas its relative is southern, their occurrence being strongly influenced by sea temperature.*

SIZE *Width to 1.5cm.*
FOOD *Filters organic matter from sea water, using its thoracic appendages (modified legs) as a net.*
SHORE ZONE *Intertidal, extending to upper shores on exposed coasts.*
DISTRIBUTION *Atlantic, south to northern Spain.*
SIMILAR SPECIES Chthamalus montagui, *which has a kite-shaped aperture;* C. stellatus, *which has an oval aperture;* Elminius modestus, *which has only four wall plates; Euraphia depressa, which has smooth plates, and is a common Mediterranean species.*

SHARES *the diverse range of intertidal habitats as its host, from intertidal rocky shores to salt marshes.*

Parasitic Barnacle

Sacculina carcini (Crustacea)

The adult Parasitic Barnacle is an obligate parasite of the Shore Crab (below) and related species. Its body has a diffuse branching structure, which invades the tissues of its host and interferes with its growth process. The only visible feature is its reproductive sac, protruding from the underside of its host.

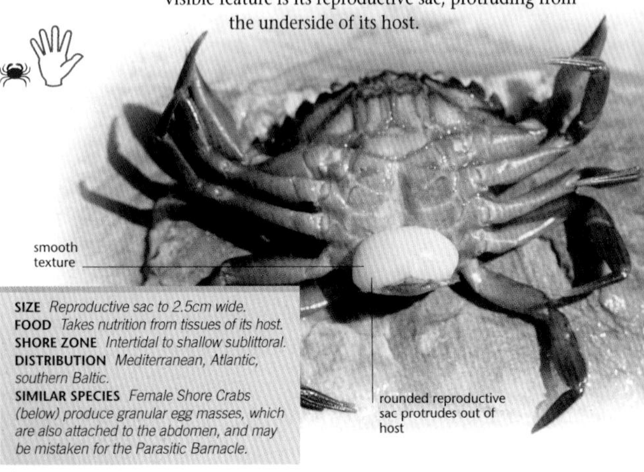

SIZE *Reproductive sac to 2.5cm wide.*
FOOD *Takes nutrition from tissues of its host.*
SHORE ZONE *Intertidal to shallow sublittoral.*
DISTRIBUTION *Mediterranean, Atlantic, southern Baltic.*
SIMILAR SPECIES *Female Shore Crabs (below) produce granular egg masses, which are also attached to the abdomen, and may be mistaken for the Parasitic Barnacle.*

ABOUNDS *in all seashore habitats, from rocky shores to salt marshes; tolerant of low salinity.*

Shore Crab

Carcinus maenas (Crustacea)

An ubiquitous crab of all shore types, the Shore Crab is variable in colour, from dark green to orange and red. Colour variation may be due to the stage of its life cycle or the habitat; juveniles, especially, display a wide range of mottled patterns. The Shore Crab scavenges on carrion and plant material, and preys upon molluscs, crustaceans, and worms.

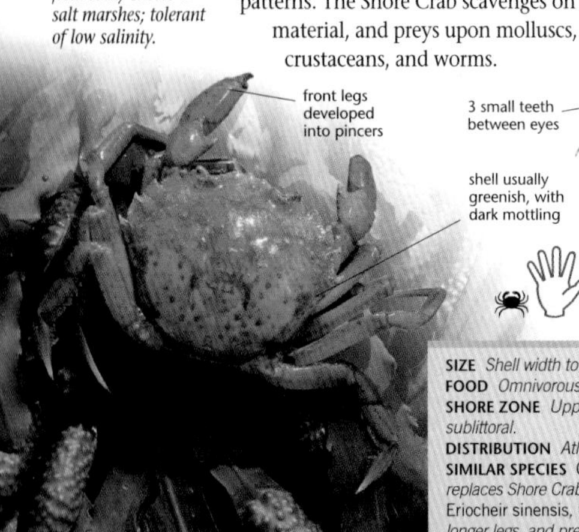

SIZE *Shell width to 8cm, length to 6cm.*
FOOD *Omnivorous.*
SHORE ZONE *Upper intertidal to shallow sublittoral.*
DISTRIBUTION *Atlantic, southern Baltic.*
SIMILAR SPECIES C. mediterraneus, *which replaces Shore Crab in the Mediterranean;* Eriocheir sinensis, *which has hairy claws and longer legs, and prefers low-salinity areas.*

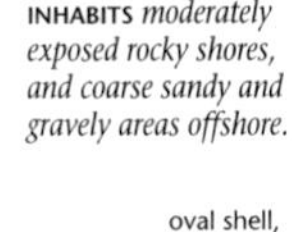

Edible Crab

Cancer pagurus (Crustacea)

Mainly found in the sublittoral zone, the female Edible Crab however moves inshore to mate and moult in spring. The Edible Crab is distinguished by its large size, massive pincers, orange-brown colour, and the indented "pie-crust" edge to its shell. It can be very long-lived, surviving beyond 20 years where stocks are not exploited.

INHABITS *moderately exposed rocky shores, and coarse sandy and gravely areas offshore.*

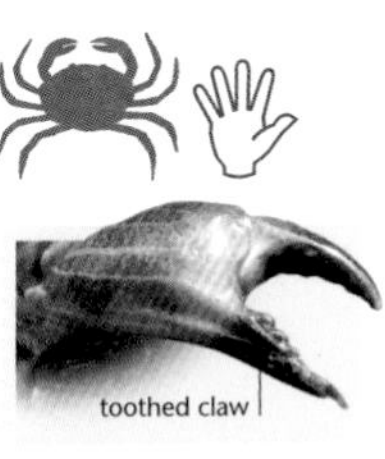

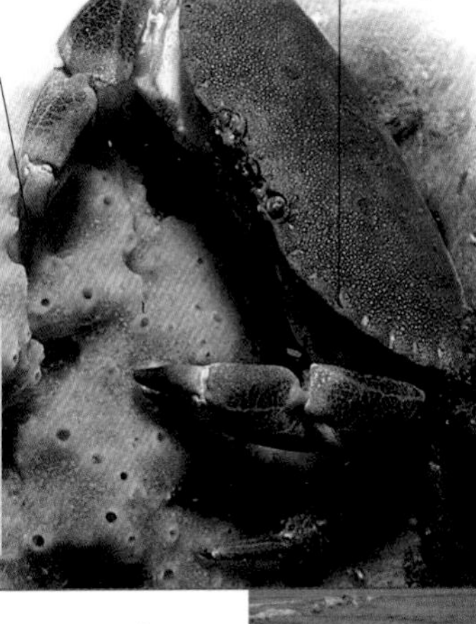

SIZE *Shell width to 25cm.*
FOOD *Preys on molluscs and crustaceans; also feeds on carrion.*
SHORE ZONE *Lower intertidal to sublittoral.*
DISTRIBUTION *Western Mediterranean, Atlantic.*
SIMILAR SPECIES Xantho incisus *is much smaller, to 4cm wide;* Pilumnus hirtellus *is even smaller, and has a hairy shell.*

Velvet Swimming Crab

Necora puber (Crustacea)

Fierce and fast-moving, the Velvet Swimming Crab, or Devil Crab, has flattened hind legs that facilitate swimming. Its shell is a dark bluish colour, often obscured by a brown, velvety coating. It bears five sharp teeth on either side of the shell, and ten small teeth between bright red eyes.

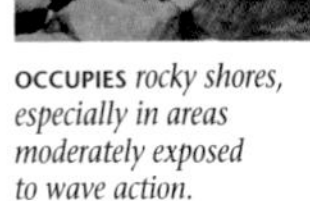

OCCUPIES *rocky shores, especially in areas moderately exposed to wave action.*

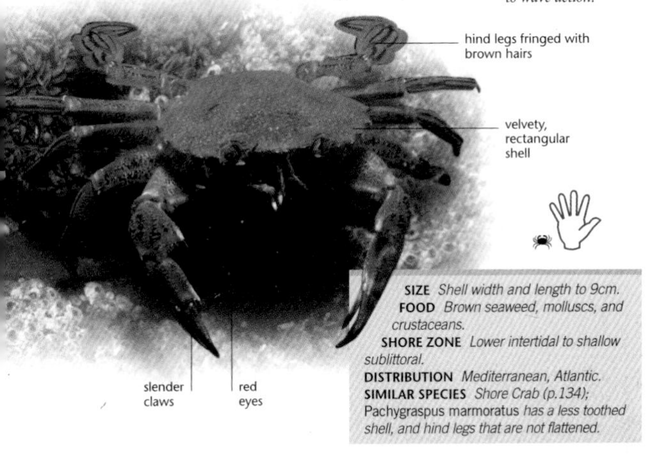

SIZE *Shell width and length to 9cm.*
FOOD *Brown seaweed, molluscs, and crustaceans.*
SHORE ZONE *Lower intertidal to shallow sublittoral.*
DISTRIBUTION *Mediterranean, Atlantic.*
SIMILAR SPECIES *Shore Crab (p.134);* Pachygraspus marmoratus *has a less toothed shell, and hind legs that are not flattened.*

Broad-clawed Porcelain-crab

HIDES *under boulders and large stones in the middle and lower intertidal zone of rocky shores.*

Porcellana platycheles (Crustacea)

A small species with a more or less circular shell, the porcelain-crabs have three pairs of walking legs, the fifth pair of legs being small and tucked under the shell. The Broad-clawed Porcelain-crab has flattened, hairy claws, and a pair of very long antennae.

SIZE *Shell width and length to 1.5cm.*
FOOD *Scavenges organic matter.*
SHORE ZONE *Intertidal to shallow sublittoral.*
DISTRIBUTION *Mediterranean; Atlantic.*
SIMILAR SPECIES Pisidia longicornis, *which is even smaller, with long, narrow claws;* Pinnotheres *species, which are smaller, to 7mm long, with soft bodies and four pairs of walking legs.*

Common Hermit-crab

INHABITS *empty mollusc shells, especially periwinkles (p.116), on intertidal gravel, and sand and rocky shores.*

Pagurus bernhardus (Crustacea)

The commonest and largest of hermit-crabs, the Common Hermit-crab has a soft abdomen that twists to fit in the whorls of a mollusc shell. Smaller, intertidal specimens usually occupy periwinkle (*Littorina*) shells, while larger crabs are found in the shells of Common Whelks (p.115).

SIZE *Head and body, when extended, to 10cm long; usually smaller intertidally.*
FOOD *Scavenges carrion, organic material.*
SHORE ZONE *Intertidal to deep sublittoral.*
DISTRIBUTION *Atlantic.*
SIMILAR SPECIES *About 20 species of hermit-crabs are found in European waters, which are distinguished by the shape of their shell and detail of their claws.*

Spiny Squat-lobster

Galathea strigosa (Crustacea)

The Squat-lobsters comprise about 10 European species, which share a common form. Their shell is narrow, with transverse ridges, and the abdomen is reflexed tightly beneath the thorax. They have a pair of forward-pointing claws and three pairs of walking legs; a fifth pair at the back is very small and slender. The Spiny Squat-lobster is the largest of Squat-lobsters, and has bright blue lines across a reddish shell.

FREQUENTS *rocky and stony shores on moderately exposed coasts, in the lower intertidal zone.*

prominent eyes

spiny rostrum

claws with sharp spines

SIZE *Shell to 5cm long.*
FOOD *Scavenges organic material from sea water.*
SHORE ZONE *Lower intertidal to deep sublittoral.*
DISTRIBUTION *Mediterranean, Atlantic.*
SIMILAR SPECIES *Many other squat-lobster species, distinguished by the shape and size of rostrum; none has bold blue markings on shell.*

Common Lobster

Homarus gammarus (Crustacea)

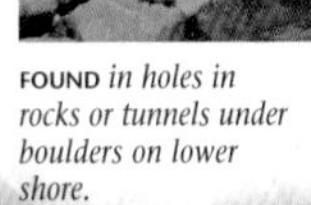

The largest European crustacean, the Common Lobster is highly prized as seafood, and therefore rarely reaches its potential maximum size. It has two large claws, which are unequal in size – the smaller one is for cutting food, the larger for crushing it. A deep bluish colour, it often has pale patches and spots, and is yellowish underneath.

FOUND *in holes in rocks or tunnels under boulders on lower shore.*

long antennae

broad tail fan

very large, unequal pincers

SIZE *Length to 1m.*
FOOD *Scavenges carrion.*
SHORE ZONE *Lower intertidal to sublittoral.*
DISTRIBUTION *Western Mediterranean, Atlantic.*
SIMILAR SPECIES Nephrops norvegicus, *which is much smaller, to 25cm, lacks massive pincers, and is found sublittorally, on sandy bottoms.*

Brown Shrimp

Crangon crangon (Crustacea)

FAVOURS *shallow water with a fine sandy or muddy bottom.*

An important commercial species, the Brown Shrimp is often buried in sediment, with only the eyes and antennae visible above the surface, in order to avoid predators and ambush prey. Usually mottled brown, it can alter its colour to suit its environment. Its antennae are almost as long as its body. It is a major food source for birds and many fish species.

SIZE *Length to 9cm.*
FOOD *Preys on small invertebrates; also scavenges plant and animal remains.*
SHORE ZONE *Intertidal to shallow sublittoral.*
DISTRIBUTION *Mediterranean, Atlantic, southern Baltic.*
SIMILAR SPECIES *Several similar shrimps, including* C. allmanni, *which has a deep central groove on the last tail segment.*

Glass Prawn

Palaemon elegans (Crustacea)

OCCURS *on rocky shores in rock pools around mid-tide level; moves offshore in winter.*

A typical prawn, the Glass Prawn is distinguished from its relatives by the shape of the rostrum between its eyes, and the teeth on its margins. It is translucent, with reddish brown lines on the shell and abdomen. The first two pairs of walking legs are tipped with small pincers, and have yellow and red banding.

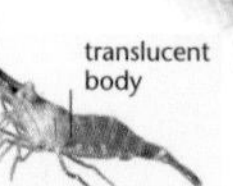

SIZE *Length to 6cm.*
FOOD *Preys on crustaceans, worms, and fish larvae; also a scavenger.*
SHORE ZONE *Intertidal, occasionally shallow sublittoral.*
DISTRIBUTION *Throughout the region.*
SIMILAR SPECIES P. longirostris, *which is larger;* P. serratus, *which is larger, with an upcurved, toothed rostrum.*

INHABITS *strandlines, among debris, especially on sheltered shores and in estuaries.*

Beach-flea

Orchestia gammarellus (Crustacea)

Like most amphipod crustaceans, the Beach-flea has large eyes and chewing mouthparts, a segmented body with no shell, two pairs of antennae, and seven pairs of legs. In this species, the second to last segment of the second pair of legs of the male is swollen into a broadly oval structure. The tail segment is short and oval, with a notch at the tip.

SIZE *Length to 1.8cm.*
FOOD *Dead seaweed and other plant materiall*
SHORE ZONE *Strandline.*
DISTRIBUTION *Throughout the region.*
SIMILAR SPECIES O. mediterranea, *which lacks the notch on the tail segment; Large Sand-hopper (below);* Gammarus *species (p.140), which are aquatic.*

Large Sand-hopper

Talitrus saltator (Crustacea)

The Large Sand-hopper is found among debris and decaying seaweed. During the day, it buries itself in sand to depths of up to 30cm. It can leap when disturbed by tucking in its tail, and rapidly flicking it out. Its body is strongly flattened along its width, and the projecting mouthparts typically hang downwards.

FOUND *on strandlines at the extreme high-water mark of sandy beaches; often extremely abundant.*

SIZE *Length to 2cm.*
FOOD *Feeds by night on decaying seaweed.*
SHORE ZONE *Strandline.*
DISTRIBUTION *Western Mediterranean, Atlantic.*
SIMILAR SPECIES Talorchestia deshayesii, *in which the male has enlarged, pincer-like tips on its second pair of legs.*

Estuarine Sand-shrimp

Gammarus duebeni (Crustacea)

FAVOURS *estuarine waters, salt marsh, and rocky shore pools; tolerant of low salinity, and in places colonizes freshwater habitats.*

Sand-shrimps in the genus *Gammarus* are relatively large amphipods which are found in aquatic habitats. They have an angular head, large kidney-shaped eyes, and long antennae. Different species favour different types of water: the Estuarine Sand-shrimp is one of the species characteristic of reduced salinity in estuaries and in the Baltic.

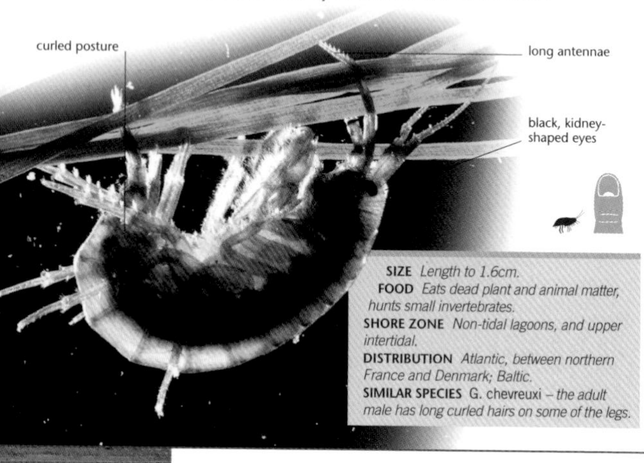

SIZE *Length to 1.6cm.*
FOOD *Eats dead plant and animal matter, hunts small invertebrates.*
SHORE ZONE *Non-tidal lagoons, and upper intertidal.*
DISTRIBUTION *Atlantic, between northern France and Denmark; Baltic.*
SIMILAR SPECIES G. chevreuxi – *the adult male has long curled hairs on some of the legs.*

Mud-shrimp

Corophium volutator (Crustacea)

LIVES *in burrows in fine mud and sand, in estuaries and lagoons; very tolerant of low salinity.*

An abundant organism of estuarine mudflats, the Mud-shrimp can occur in densities of up to 100,000 per square metre. It occupies a U-shaped tube in muddy sediments, across a wide range of salinity. It is an important source of food for birds such as Dunlin (p.193), Redshank (p.191), and Shelduck (p.185), and fish such as Plaice (p.160).

SIZE *Length to 1cm.*
FOOD *Feeds on deposited and floating organic material.*
SHORE ZONE *Intertidal.*
DISTRIBUTION *Throughout the region.*
SIMILAR SPECIES C. arenarium, *which is smaller, and replaces* C. volutator *in sandier sediments and saltier water; and* C. insidiosum, *which lives in silty tubes.*

Sea-slater

Ligia oceanica (Crustacea)

A large number of isopods occur in the marine environment, many of which resemble their terrestrial relatives, the woodlice. The largest coastal species is the Sea-slater, to 3cm long, which occurs around the high-water mark of all rocky habitats. Like all isopods, it has seven pairs of legs. The Sea-slater lives for up to three years, but usually breeds only once.

CRAWLS *over all manner of rocky coasts, including harbour walls, above and around the high-water mark.*

SIZE *Length to 3cm.*
FOOD *Eats dead plant material, and grazes encrusting microscopic algae.*
SHORE ZONE *High-water mark upwards.*
DISTRIBUTION *Mediterranean, Atlantic, western Baltic.*
SIMILAR SPECIES L. italica *(Mediterranean only), which is smaller, with longer rear projections.*

Janira maculosa

Janira maculosa (Crustacea)

One of many small isopods, of which there are some 200 marine species in Europe, *Janira maculosa* has the appearance of a small, narrow woodlouse, with long antennae. It has a greyish brown colour, and the plates covering the segments appear rough due to minute sculpturing. Given the number of related species, precise identification of this group is difficult, and best left to the expert.

OCCURS *at the lower intertidal levels of rocky shores, often among seaweed holdfasts or encrusting sponges.*

2 rear projections

narrowly oval body, with almost parallel sides

1 pair of antennae much longer than the other pair

SIZE *Length to 1cm.*
FOOD *Eats organic debris.*
SHORE ZONE *Lower intertidal and shallow sublittoral.*
DISTRIBUTION *Atlantic, north from Biscay.*
SIMILAR SPECIES Jaera *species, which are more oval, with visible legs;* Limnoria *species, which have shorter antennae, and live as borers of wet wood.*

Rock Springtail

Anurida maritima (Collembola)

FOUND *on intertidal rocky shores, often collecting on the surface of rock pools, retreating into crevices as the tide comes in.*

Springtails are primitive wingless insects, and one of the groups primarily responsible for the decay of organic matter. The Rock Springtail is one of a few air-breathing insects capable of immersion in salt water. When covered by tide, the body hairs trap a layer of air, which it can breathe. Unlike most springtails, it lacks a functional springing organ at the rear of its abdomen.

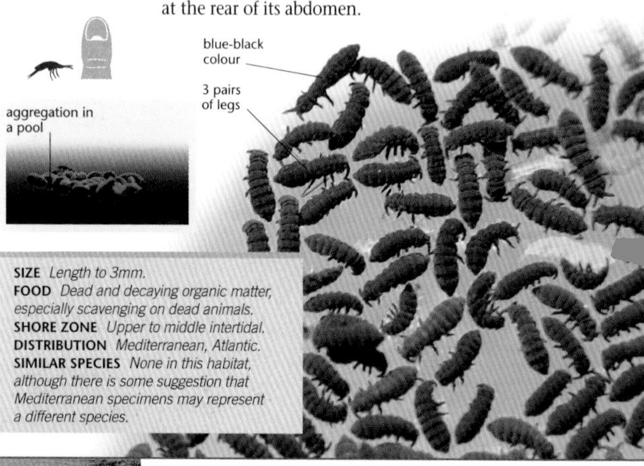

SIZE *Length to 3mm.*
FOOD *Dead and decaying organic matter, especially scavenging on dead animals.*
SHORE ZONE *Upper to middle intertidal.*
DISTRIBUTION *Mediterranean, Atlantic.*
SIMILAR SPECIES *None in this habitat, although there is some suggestion that Mediterranean specimens may represent a different species.*

Sea Bristletail

Petrobius maritimus (Archaeognatha)

THRIVES *among coastal rocks, above the high tide level.*

Similar in appearance to the house-dwelling Silverfish (*Lepisma saccharina*), the Sea Bristletail has a long, tapering abdomen, with three tails, and a pair of segmented antennae, which are as long as or longer than the body. It is one of the jumping bristletails – its tail helps to propel it several centimetres into the air. It runs swiftly with a scuttling motion, and is most easily seen on rocks at night by torchlight.

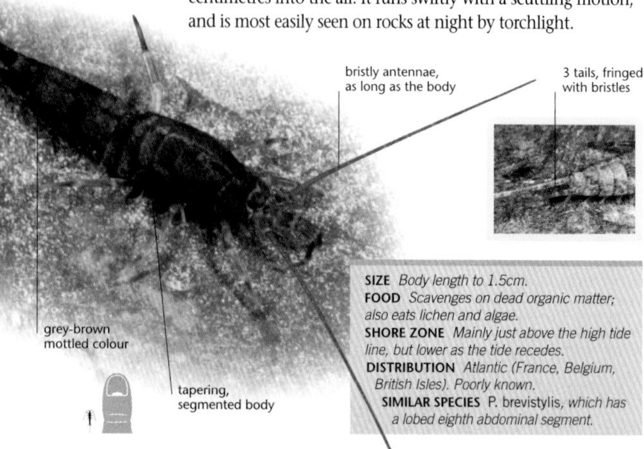

SIZE *Body length to 1.5cm.*
FOOD *Scavenges on dead organic matter; also eats lichen and algae.*
SHORE ZONE *Mainly just above the high tide line, but lower as the tide recedes.*
DISTRIBUTION *Atlantic (France, Belgium, British Isles). Poorly known.*
SIMILAR SPECIES P. brevistylis, *which has a lobed eighth abdominal segment.*

Kelp-fly

Coelopa frigida (Diptera)

Swarming on seaweed-covered strandlines, Kelp-flies are one of the most numerous insects to be found on the beach. Although most abundant during summer, they can be seen throughout the year in warm weather. A closely related species, *C. pilipes*, which has a more northerly distribution, is distinguished by being much more hairy, especially the male.

BREEDS *in rotting seaweed on rocky and other shorelines; adults feed during the summer on coastal flowers.*

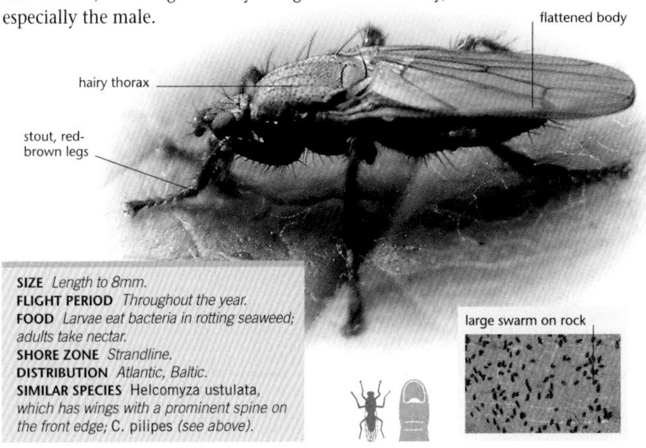

SIZE *Length to 8mm.*
FLIGHT PERIOD *Throughout the year.*
FOOD *Larvae eat bacteria in rotting seaweed; adults take nectar.*
SHORE ZONE *Strandline.*
DISTRIBUTION *Atlantic, Baltic.*
SIMILAR SPECIES Helcomyza ustulata, *which has wings with a prominent spine on the front edge;* C. pilipes *(see above).*

Flecked General

Stratiomys singularior (Diptera)

One of a group of flies called soldier-flies, on account of their usually bold, sometimes metallic colours, the Flecked General is characteristic of brackish coastal marshes. Females have broader bodies than the males, and two narrow yellow bars on the head, just above the antennae. The males have eyes fringed with hair. The larvae are long-lived, and their long body tapers to a fine point.

INHABITS *brackish habitats, where larvae feed for up to five years in muddy ditches.*

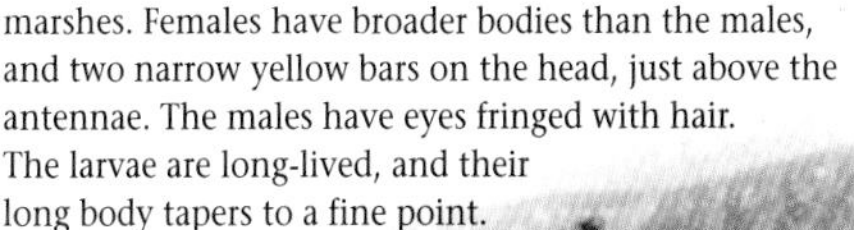

SIZE *Adults 1.2–1.5cm long; larvae to 6cm long.*
FOOD *Adults take nectar; larvae feed on decaying organic matter.*
FLIGHT PERIOD *May–September.*
DISTRIBUTION *Western Mediterranean, Atlantic, southern Baltic.*
SIMILAR SPECIES S. longicornis, *found in even more saline waters, is almost unmarked.*

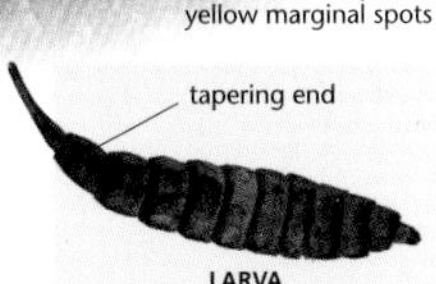

LARVA

Red-banded Sand-wasp

Ammophila sabulosa (Hymenoptera)

FAVOURS *sand dunes and other sandy habitats, both coastal and inland. Nests in bare ground.*

An active, solitary digger wasp, the Red-banded Sand-wasp is often seen darting across bare sand. Nest holes are excavated in bare sand by the female, using its comb-like front legs. The holes are provisioned with caterpillars that have been paralysed by stinging, before being carried or dragged to the nest. Eggs are laid on the caterpillars and the burrow is closed with a small stone.

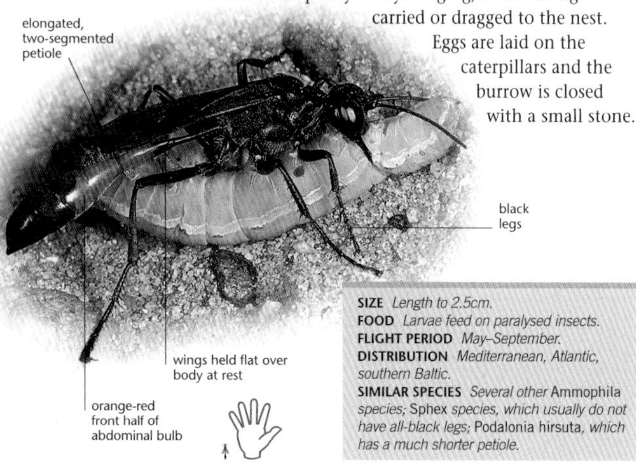

SIZE *Length to 2.5cm.*
FOOD *Larvae feed on paralysed insects.*
FLIGHT PERIOD *May–September.*
DISTRIBUTION *Mediterranean, Atlantic, southern Baltic.*
SIMILAR SPECIES *Several other* Ammophila *species;* Sphex *species, which usually do not have all-black legs;* Podalonia hirsuta, *which has a much shorter petiole.*

Saltmarsh Mining-bee

Colletes halophilus (Hymenoptera)

NESTS *in the upper salt marsh zone and immediate terrestrial fringes; forages for pollen and nectar on late summer flowers.*

The Saltmarsh Mining-bee often nests in huge aggregations on bare ground, including low, sandy banks next to footpaths, where each nest is excavated by a single female. In appearance rather like a small Honey-bee, the *Colletes* species are difficult to distinguish, but the Saltmarsh Mining-bee has distinct habitat preferences and a late flight period. Pollen is carried to the nest on the hairs of the legs.

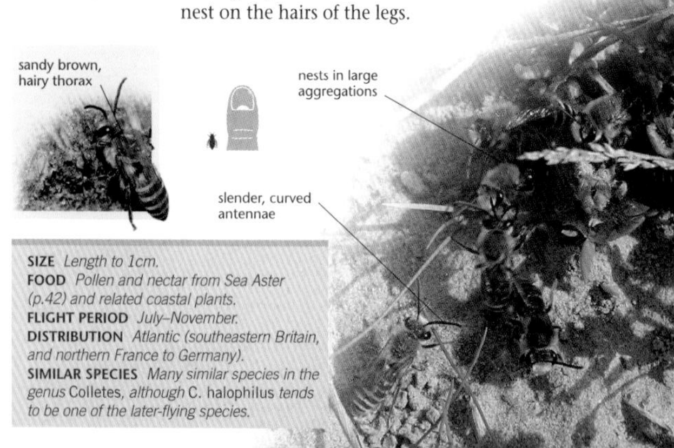

SIZE *Length to 1cm.*
FOOD *Pollen and nectar from Sea Aster (p.42) and related coastal plants.*
FLIGHT PERIOD *July–November.*
DISTRIBUTION *Atlantic (southeastern Britain, and northern France to Germany).*
SIMILAR SPECIES *Many similar species in the genus* Colletes, *although* C. halophilus *tends to be one of the later-flying species.*

Black Oil-beetle

Meloe proscarabaeus (Coleoptera)

PREFERS *warm, open habitats, especially coastal in the north of its range, with the nests of solitary bees.*

Oil-beetles have a fascinating life cycle. In spring, the female digs a burrow, in which she lays around 1,000 eggs. These hatch the following year, and the active larvae (tringulins) then look for a host. The tringulin attaches itself to a bee when it is visiting flowers, and travels to the bee's nest, where it turns into a grub-like larva, feeding on the pollen stores and eggs of the host. Black Oil-beetle is the most widespread of the larger species, and releases an oily, smelly fluid when threatened.

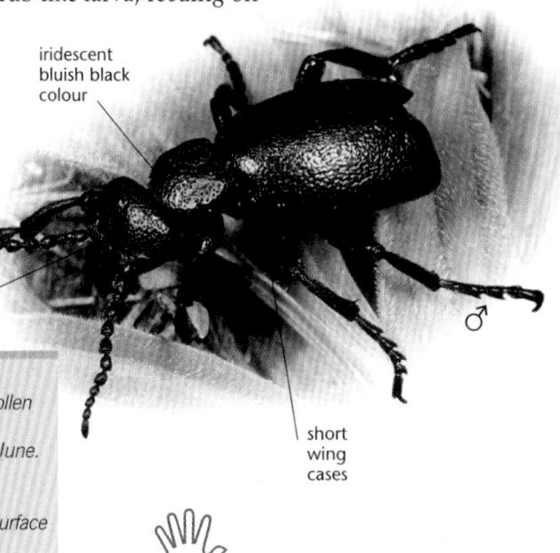

SIZE *Length to 3.5cm.*
FOOD *Adults chew leaves; larvae eat pollen and eggs in nests of solitary bees.*
FLIGHT PERIOD *Adults usually March–June.*
DISTRIBUTION *Throughout the region.*
SIMILAR SPECIES *Several other species, differing in colour and the detail of their surface sculpturing, such as* M. violaceus, *which usually has a strong blue sheen.*

Green Tiger-beetle

Cicindela campestris (Coleoptera)

THRIVES *in warm, sandy habitats, including sand dunes, soft eroding cliff slopes, and heathland.*

A warmth-loving, fast-running, active predator of invertebrates, the long legs, big eyes, and fearsome mouthparts of the adult Green Tiger-beetle make it well-equipped for hunting. Larvae are similarly endowed with powerful jaws, which they use to ambush passing prey from the shelter of their burrows. The adult readily takes flight, making a loud, buzzing sound. It is bright metallic green above and often has a coppery, purple sheen.

SIZE *Length to 1.5cm.*
FOOD *Invertebrates; adults hunt, while larvae ambush their prey from burrows. The prey is held in the jaws while digestive juices are secreted onto it.*
FLIGHT PERIOD *April–August.*
DISTRIBUTION *Throughout the region.*
SIMILAR SPECIES C. hybrida, *which is tinged with red and has more extensive yellow marks.*

Grayling

Hipparchia semele (Lepidoptera)

FAVOURS *dry, sunny habitats, including sand dunes and cliff-tops; on dry grassland and mountains inland.*

A sun-loving butterfly, the Grayling is usually found settled on bare, sandy ground or rocks, with its wings closed. Its underside markings provide effective camouflage against such backgrounds, enhanced by its habit of tilting towards the sun so as to cast the minimum shadow. The brown, variably-spotted upperwings are rarely revealed to observers.

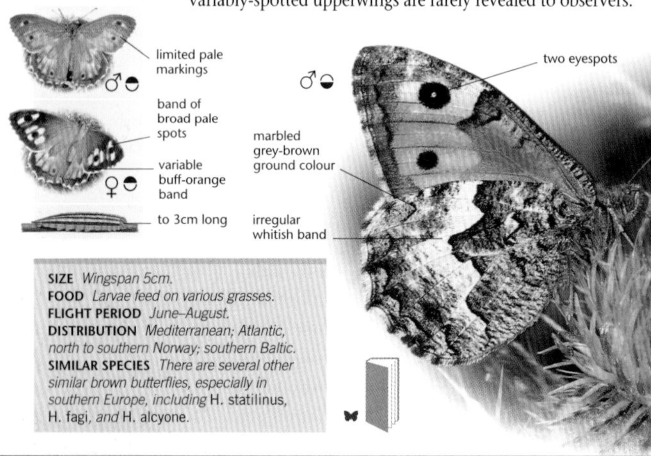

SIZE *Wingspan 5cm.*
FOOD *Larvae feed on various grasses.*
FLIGHT PERIOD *June–August.*
DISTRIBUTION *Mediterranean; Atlantic, north to southern Norway; southern Baltic.*
SIMILAR SPECIES *There are several other similar brown butterflies, especially in southern Europe, including* H. statilinus, H. fagi, *and* H. alcyone.

Ground Lackey

Malacosoma castrensis (Lepidoptera)

PREFERS *upper salt marshes and shingle banks in the north of its range; more widespread further south, including inland.*

Although the adult Ground Lackey is rarely seen unless specifically searched for at night, the caterpillars of this moth are a familiar sight on many salt marshes. The larvae live communally, spinning silken webs, within which they can retreat if the marshes are inundated by an unexpectedly high tide. In sunshine, they often bask openly, protected from most predators by their long, dense, irritant hairs.

SIZE *Wingspan to 4.5cm.*
FOOD *Larvae feed on a wide range of low-growing plants.*
FLIGHT PERIOD *July–September.*
DISTRIBUTION *Mediterranean; Atlantic, north to southeast Britain and southern Norway; southern Baltic.*
SIMILAR SPECIES M. neustria, *which has more parallel cross-lines.*

Grey Bush-cricket

Platycleis albopunctata (Orthoptera)

INHABITS *dry grassy habitats, including sand dunes and soft cliff slopes; in the north, usually within sight of the sea.*

Bush-crickets are related to grasshoppers, but differ in their elongated antennae and partly carnivorous diet. The female lays eggs in plant stems or crevices, using its long, curved ovipositor. The Grey Bush-cricket is a medium-sized species, with a generally grey or dull brown colour. A warmth-loving insect, it is wary, readily taking flight when disturbed.

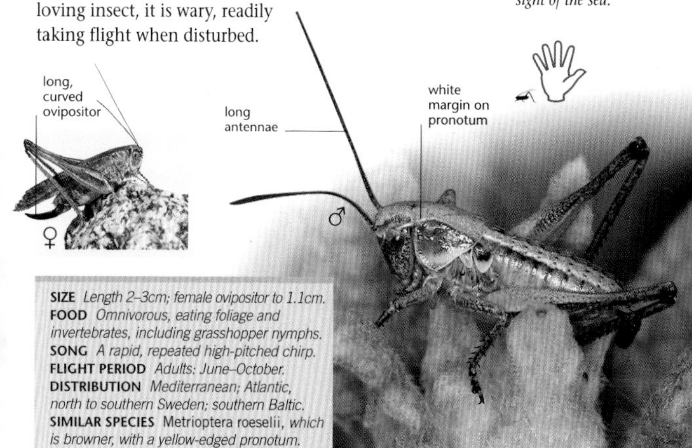

SIZE *Length 2–3cm; female ovipositor to 1.1cm.*
FOOD *Omnivorous, eating foliage and invertebrates, including grasshopper nymphs.*
SONG *A rapid, repeated high-pitched chirp.*
FLIGHT PERIOD *Adults: June–October.*
DISTRIBUTION *Mediterranean; Atlantic, north to southern Sweden; southern Baltic.*
SIMILAR SPECIES Metrioptera roeselii, *which is browner, with a yellow-edged pronotum.*

Nosed Grasshopper

Acrida ungarica (Orthoptera)

OCCURS *in coastal marshland and grassland, and on open sand dunes; also found inland.*

An elongated, slender grasshopper, the Nosed Grasshopper has an unmistakable shape, at least within its family. It is found in two colour forms. The green form is usually found in more lush, vegetated habitats, whereas the pale brown form, which looks like a bundle of dead grass stems, inhabits dry, sandy sites. Its very long, slender legs accentuate its straw-like appearance.

SIZE *Hind leg to 7cm long.*
FOOD *Leaves and other vegetable matter.*
SONG *None.*
FLIGHT PERIOD *Adult: May–October.*
DISTRIBUTION *Mediterranean.*
SIMILAR SPECIES Saga pedo, *a large, slender, green bush-cricket, which is wingless; some unrelated stick-insects and praying mantises.*

European Ant-lion

Euroleon nostras (Neuroptera)

INHABITS *dry, sandy areas, predominantly coastal, especially in the north of its range; isolated trees are important as mating sites.*

Named after the feeding habits of the larvae, which ambush ground-dwelling invertebrates that stumble into their pit, the adult European Ant-lion looks similar to a damselfly, though the clubbed antennae are quite different. The male has claspers at the end of the abdomen; the larvae are short and sac-like, with huge jaws.

LARVAL PIT

SIZE *Length to 3cm; wingspan to 7cm.*
FOOD *Adults may not feed at all, or pluck insects; larvae eat invertebrates, including egg-laying adult female ant-lions.*
FLIGHT PERIOD *July–August.*
DISTRIBUTION *Central Mediterranean; Atlantic, France to Denmark; southern Baltic.*
SIMILAR SPECIES Acanthoclisis baetica, *which is a larger, unspotted ant-lion.*

Scarce Emerald

Lestes dryas (Odonata)

OCCUPIES *shallow coastal standing waters, including brackish and temporary sites; also inland wetlands.*

One of the more widespread species of seven emerald damselflies in Europe, the Scarce Emerald can be identified by the shape and colour of the pterostigma, a coloured panel near the wingtips, and by the distribution of the blue colour on the male's abdomen. Overwintering as eggs, followed by rapid nymphal development, it allows to occupy temporary pools, unsuitable for its predators.

SIZE *Length to 4cm; wingspan to 5.5cm.*
FOOD *Adults and nymphs prey on small invertebrates, picked off vegetation or captured underwater, respectively.*
FLIGHT PERIOD *April–October.*
DISTIBUTION *Eastern Mediterranean; Atlantic, north to southern Sweden; Baltic.*
SIMILAR SPECIES *Other* Lestes *species;* Calopteryx *species, which are metallic green.*

Red-veined Darter

Sympetrum fonscolombei (Odonata)

Like most darter dragonflies, male Red-veined Darters are bright red, and females are yellow-brown on the abdomen. The most distinctive feature is the coloured veins in the front part of each wing. A dispersive species, often found well away from its core range, it is also strongly territorial, returning to the same prominent perches on twigs and vegetation after feeding forays.

FAVOURS *shallow pools, including coastal lagoons and dune slacks; often in brackish and temporary pools.*

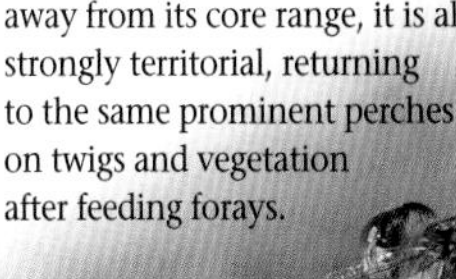

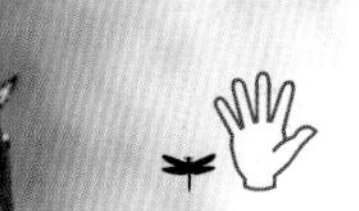

SIZE *Length 3.5–4cm; wingspan 6.5cm.*
FOOD *Adults and nymphs catch small invertebrates in flight or underwater, respectively.*
FLIGHT PERIOD *All year in the south, but particularly March–October.*
DISTRIBUTION *Mediterranean; Atlantic, north to Brittany; dispersive north to Baltic.*
SIMILAR SPECIES *Scarlet Darter (below).*

Scarlet Darter

Crocothemis erythraea (Odonata)

The male Scarlet Darter is bright scarlet in colour, lacking any dark markings. It has a darting flight, between perches on vegetation and the ground. Like all Odonata species, the colour takes several days to develop; however, the broad, strongly flattened body is distinctive even in immature individuals. Once restricted to the Mediterranean, it is now readily colonizing more northern areas.

OCCURS *in brackish lagoons and other warm, shallow coastal waters; relatively tolerant of pollution; also found inland.*

SIZE *Length 3.5-4.5cm; wingspan 7cm.*
FOOD *Adults and nymphs prey on small invertebrates, captured in flight or underwater.*
FLIGHT PERIOD *March–October.*
DISTRIBUTION *Mediterranean; Atlantic, north to Belgium; currently spreading northwards.*
SIMILAR SPECIES *Red-veined Darter (above);* Sympetrium sanguineum, *which is redder, with a narrow, waisted abdomen.*

Green-eyed Hawker

Aeshna isosceles (Odonata)

FAVOURS *coastal grazing marsh ditches and lagoons with clear, unpolluted water.*

Unlike many dragonflies, the sexes of the Green-eyed Hawker (also known as Norfolk Hawker on account of its restricted British distribution) are very similar. A medium-sized, mid-brown hawker, its most obvious features are its large, green eyes, and the yellow triangle on its second abdominal segment. Its scientific name *isosceles* is derived from this feature.

green eyes

yellow triangle on second abdominal segment

SIZE *Length to 6.5cm; wingspan 8–9cm.*
FOOD *Adults and nymphs catch small invertebrates in flight and in water, respectively.*
FLIGHT PERIOD *May–August.*
DISTRIBUTION *Mediterranean; Atlantic, from France to Britain and Denmark; southern Baltic.*
SIMILAR SPECIES A. grandis, *which is larger, darker, and has brown wing membranes.*

Hairy Dragonfly

Brachytron pratense (Odonata)

INHABITS *well-vegetated ditches in coastal grazing marshes.*

A small, secretive hawker, the adult Hairy Dragonfly spends much of its time flying among emergent vegetation, patrolling in search of food. The hairy thorax is distinctive; in the male it is green, with two narrow, black stripes, while in the female it is brown, sometimes with restricted yellow marks.

dense hair on thorax

pale spot on first abdominal segment

♀ paired yellow spots on abdomen

♂

SIZE *Length 5.5–6.5cm; wingspan 7–8cm.*
FOOD *Adults and nymphs prey on small invertebrates, captured in flight or underwater, respectively.*
FLIGHT PERIOD *March–May.*
DISTRIBUTION *Central Mediterranean; Atlantic, Spain to southern Norway; southern Baltic.*
SIMILAR SPECIES Aeshna cyanea *males have apple-green markings on abdomen and thorax.*

Distinguished Jumper

Sitticus distinguendus (Arachnida)

FAVOURS *vegetated sandy inland sites; also a colonist of coastal brownfield sites.*

As active predators that jump on their prey from a distance of several centimetres, jumping spiders require acute eyesight. They have four pairs of eyes: one large pair (like headlamps) and one small pair at the front of the carapace, and two more pairs further back to give an all-round vision. The Distinguished Jumper is a greyish, frosted species of sand dunes and similar habitats. It has a variable pattern of dark and pale hairs on the back and legs.

SIZE *Length to 7mm.*
FOOD *Preys on small invertebrates.*
ADULT PERIOD *April–September.*
DISTRIBUTION *Mediterranean; Atlantic, France to Norway; Baltic.*
SIMILAR SPECIES *Other* Sitticus *species – males can be distinguished by the pattern of hair on the face, but females can only be differentiated by microscopic examination.*

pale, frosted colouring

Jumping Spider

Heliophanus auratus (Arachnida)

FREQUENTS *shingle and shell banks, in sparse vegetation and litter, often around the spring tide line; also found inland.*

A small, blackish jumping spider, *Heliophanus auratus* is characteristic of shingle banks, and may often be seen running over the stones in warm weather. It usually has pale hairs which contrast with the black body; these hairs may form distinct spots or chevrons, or spread across the whole abdomen, giving it a greenish golden, metallic appearance.

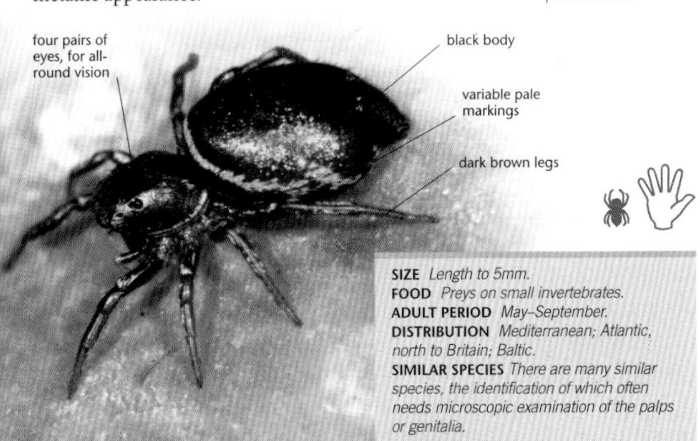

SIZE *Length to 5mm.*
FOOD *Preys on small invertebrates.*
ADULT PERIOD *May–September.*
DISTRIBUTION *Mediterranean; Atlantic, north to Britain; Baltic.*
SIMILAR SPECIES *There are many similar species, the identification of which often needs microscopic examination of the palps or genitalia.*

Vertebrates

Covering all vertebrate (backboned) groups apart from birds, this section encompasses mammals, reptiles, amphibians, and fish. The fish species included here are mostly those that are found in water-filled intertidal habitats above the low-water mark, such as rock pools, or in very shallow waters, and so can be seen by the land-based observer. Some mammals (whales, dolphins, and seals) and reptiles (sea-turtles) are also exclusively or primarily marine, but many more (and all amphibians, such as frogs and toads) are found above the reach of tidal waters.

RINGED SEAL

SHANNY

LOGGERHEAD TURTLE

NATTERJACK TOAD

Otter

Lutra lutra (Mammalia)

A predominantly aquatic mammal, the Otter is generally seen in or near water. However, it will travel long distances overland at night, its main activity period: at such times, otters are frequently killed on roads. On land, it moves with an awkward, bounding gait, but in water it is very agile and playful. It swims low in the water, with just the head exposed, lower than any of the minks or large aquatic rodents with which it may be confused. When underwater, it has a silvery appearance due to the bubbles of air trapped in the fur. It has whitish underparts and small ears.

FAVOURS *estuaries, lagoons, and sheltered rocky coasts, without significant disturbance or pollution; also inland waters.*

NOTE

Especially in northern Europe, Otters can be found on rocky shores covered with seaweed and adjacent shallow waters, where crabs and molluscs form the major part of their diet.

SIZE *Body 55–90cm; tail 35–50cm.*
YOUNG *Single litter of 2–5; May–August, but can be all year round.*
DIET *Fish, amphibians, rodents, water birds, and crustaceans; crabs and molluscs in coastal areas; carrion.*
DISTRIBUTION *Throughout the region.*
SIMILAR SPECIES *Both American Mink* (Mustela vison) *and European Mink* (M. lutreola) *can be found in coastal areas; they are smaller and darker than the Otter, and the European Mink also has a white chin patch.*

Common Seal

Phoca vitulina (Pinnipeda)

OCCUPIES *shallow coastal waters and estuaries, sometimes swimming great distances up rivers; seen on rocks and sand banks at low tide.*

Also known as the Harbour Seal, the Common Seal is one of the smaller members of its family. Its short muzzle protrudes from the forehead, giving it a dog-like profile. Pups are born with a colour similar to the parents, and are capable of swimming almost immediately after birth. Common Seals can, therefore, breed on tidal flats.

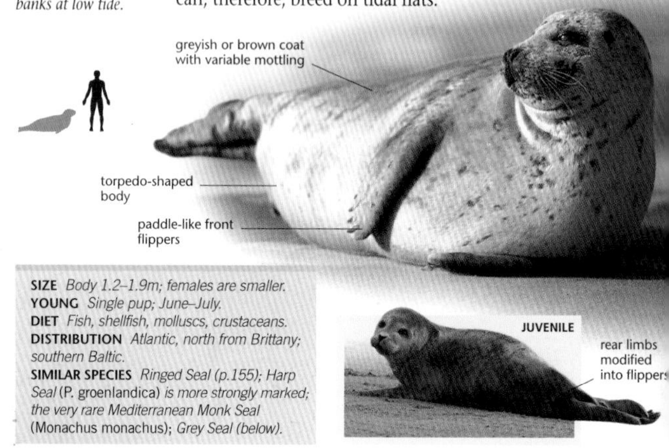

SIZE *Body 1.2–1.9m; females are smaller.*
YOUNG *Single pup; June–July.*
DIET *Fish, shellfish, molluscs, crustaceans.*
DISTRIBUTION *Atlantic, north from Brittany; southern Baltic.*
SIMILAR SPECIES *Ringed Seal (p.155); Harp Seal* (P. groenlandica) *is more strongly marked; the very rare Mediterranean Monk Seal* (Monachus monachus); *Grey Seal (below).*

JUVENILE

rear limbs modified into flippers

Grey Seal

Halichoerus grypus (Pinnipeda)

FOUND *along coasts and in coastal marine waters, breeding on rocky islets, grassy coastal strips, and ice shelves in the north.*

Male Grey Seals are very large, about a third larger than the females. Both sexes have variably blotchy grey upperparts and paler undersides. In profile, the forehead runs straight into the muzzle and the nostrils are widely separated. Pups are born white and remain on land for several weeks. Grey Seals, therefore, must breed above the high tide mark. In water, they are capable of diving to depths of more than 200m, remaining submerged for up to 30 minutes at a time.

♂

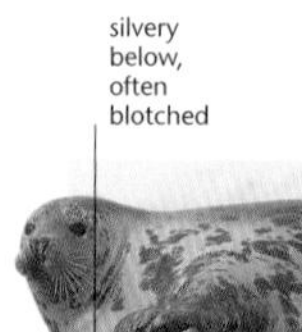

Ringed Seal

Phoca hispida (Pinnipeda)

The smallest European seal, with males and females similar in size, the Ringed Seal usually has a dark grey-brown back. Like the Common Seal, it has a dog-like, concave facial profile. It is the most common seal found in the Arctic, breeding also in the Baltic, and rarely wandering far from its breeding areas; as a result it is rarely or never seen around most of the European coastline. Pups are white and woolly at birth.

SWIMS *in inshore water, bays, and fjords, and breeds on ice sheets; readily enters fresh water.*

SIZE *Body 1.2–1.35m.*
YOUNG *Single litter of 1–2; February–April.*
DIET *Fish and small crustaceans.*
DISTRIBUTION *Baltic.*
SIMILAR SPECIES *Common Seal (p.154), which is a little larger, and has less prominent markings; Grey Seal (below), which is often blotchy, with an elongate muzzle.*

SIZE *Males 2.2–3m; females 2–2.5m.*
YOUNG *Single pup; normally October–February.*
DIET *Fish (including commercial species such as salmon), crustaceans, and cephalopods.*
DISTRIBUTION *Atlantic, north from Brittany; Baltic.*
SIMILAR SPECIES *Common and Ringed Seals (p.154), both of which are smaller and have snub noses; Walrus* (Odobaenus rosmarus)*, which is larger with prominent tusks and a characteristic upright stance on land, and is found on Norwegian coasts.*

NOTE

The long, straight facial profile of this seal is especially characteristic of males; females and juveniles have a shorter muzzle, but never show the concave profile of most other European seal species.

Harbour Porpoise

Phocoena phocoena (Cetacea)

OCCURS *in shallow coastal waters and estuaries; substantial aggregations can form in favoured feeding areas, although declining in some areas due to disturbance.*

The commonest (and smallest) cetacean in European waters, the Harbour Porpoise is relatively nondescript – it is steely grey above and whitish below, the pigmentation usually being asymmetric. The dorsal fin, often the only feature visible above water, is short and blunt, and located centrally down the back. This porpoise has a rounded head and spade-shaped teeth. Unlike most dolphins, its snout is not extended into a beak. It is not as agile as a dolphin and does not leap clear of the water as a dolphin does.

NOTE

The likelihood of viewing small cetaceans such as Harbour Porpoises depends upon weather conditions; the merest hint of waves easily disguises the occasional fleeting dorsal fin.

Bottlenose Dolphin

Tursiops truncatus (Cetacea)

FOUND *worldwide in tropical and temperate seas, from shallow estuaries to deep ocean waters.*

The most widespread and abundant dolphin over much of the European coastline, the Bottle-nosed Dolphin has a short beak, greyish upperparts, and a long, curved dorsal fin. Despite its bulky appearance, it is very acrobatic, often leaping clear of the water. It can be very inquisitive, approaching swimmers and boats closely.

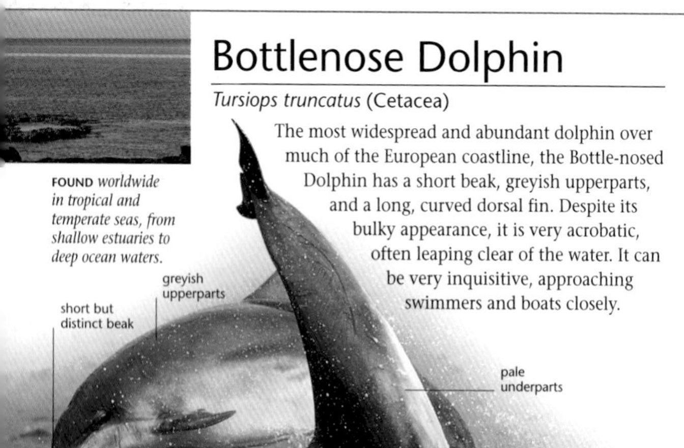

SIZE *2.5–4m.*
YOUNG *Single calf; born usually April–September, at intervals of up to three years.*
DIET *Fish, cuttlefish, and squid.*
DISTRIBUTION *Atlantic, Mediterranean.*
SIMILAR SPECIES *Risso's Dolphin* (Grampus griseus) *and Long-finned Pilot-whale* (Globicephala melas), *which are pale grey and blackish, respectively.*

SIZE *1.4–1.8m.*
YOUNG *Single young; born May–August; not every year.*
DIET *Small fish, especially herrings; also some crustaceans and cuttlefish.*
DISTRIBUTION *Atlantic, rare and threatened populations in the Mediterranean and Baltic.*
SIMILAR SPECIES *No small European cetacean has the blunt head and small, triangular fin, although the much larger Minke Whale* (Balaenoptera acutorostrata) *can have a similar fin shape.*

Basking Shark

Cetorhinus maximus (Pisces)

Despite its size (it is the second largest fish in the world) and fearsome gaping mouth, the Basking Shark is a harmless filter-feeder, trawling through inshore waters, collecting plankton. The first, and often only, sign of its presence is the large triangular dorsal fin, followed by the tip of its tail fin, moving from side to side as it slowly swims forward.

INHABITS *the open sea, moving into shallow coastal waters, especially during the summer months.*

SIZE *To 13m long.*
YOUNG *Up to six young, born after a gestation of 2–3 years.*
DIET *Plankton.*
DISTRIBUTION *Mediterranean, Atlantic.*
SIMILAR SPECIES *Other large sharks occur in the region, but usually in deeper waters; Killer Whale* (Orcinus orca) *has an even larger dorsal fin (to 1.8m high), and is black and white.*

Maned Sea-horse

Hippocampus guttulatus (Synghathidae)

INHABITS *beds of seaweed and sea-grasses around and below the low-water mark.*

Sea-horses have several unique features, including a horse-like head profile, an upright swimming posture, a prehensile tail, and ribbed body. The Maned Sea-horse can be distinguished from similar species by the mane-like protuberances from the back of its head and neck.

fat body

long, straight, tubular snout

mane of fleshy projections

tapering, prehensile tail

SIZE *Length to 15cm.*
DIET *Takes small planktonic animals.*
BREEDING *April–October; eggs brooded by the male in a pouch.*
DISTRIBUTION *Mediterranean; Atlantic, north to southern Britain.*
SIMILAR SPECIES *Short-snouted Sea-horse* (H. hippocampus), *which has a short, upturned snout, and lacks the mane.*

Lesser Sand-eel

Ammodytes tobianus (Ammodytidae)

FOUND *on sandy shores, from mid-tide level to the shallow sublittoral; often buried in sand.*

A shoaling, silvery fish, the Lesser Sand-eel is often an abundant species in the inshore zone. If threatened, it dives down and disappears into the sand. It has a long, thin body with a pointed jaw, and a single, long dorsal fin. It forms the staple diet of many seabirds, including the Puffin (p.179) and terns (pp.176–77). However, many stocks have been fished on an industrial scale, and this has greatly affected the productivity of birds that rely on sand-eels to feed their chicks.

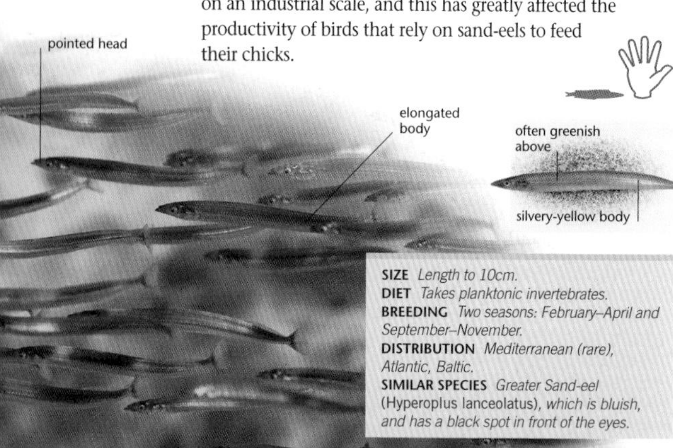

often greenish above

silvery-yellow body

SIZE *Length to 10cm.*
DIET *Takes planktonic invertebrates.*
BREEDING *Two seasons: February–April and September–November.*
DISTRIBUTION *Mediterranean (rare), Atlantic, Baltic.*
SIMILAR SPECIES *Greater Sand-eel* (Hyperoplus lanceolatus), *which is bluish, and has a black spot in front of the eyes.*

Lesser Weever-fish

Echiichthys vipera (Trachinidae)

The scourge of many a seaside holiday, the Lesser Weever-fish, and others in its family, have poison glands associated with spines on the first dorsal fin and gill-covers. Buried in soft sand and mud, it frequently stings bathers. The Lesser Weever-fish has a distinctive wedge-shaped appearance, with its eyes on top of its head, and an upward-sloping mouth.

LIVES *in shallow water and pools on sandy shores, buried in the sediment so that only the eyes and dorsal fins are exposed.*

SIZE *Length to 15cm.*
DIET *Takes crustaceans and small fish.*
BREEDING *June–August.*
DISTRIBUTION *Mediterranean; Atlantic, north to Scotland and Denmark.*
SIMILAR SPECIES *Greater Weever* (Trachinus draco), *which is larger (to 40cm long), has a concave tail fin, and often a greenish tinge to the body.*

Shore Clingfish

Lepadogaster lepadogaster (Gobiesocidae)

The Shore Clingfish has a flattened body with a broad triangular head, and a "duck-billed" snout accentuated by thick lips. The pelvic fins are modified into a thoracic sucking disc that allows it to adhere to rocks and other substrates. Usually a reddish brown colour with darker smudges, its most obvious markings are two dark blue spots behind the eyes.

OCCUPIES *seaweed-covered rocky shores and rock pools; clings to the underside of boulders at low tide.*

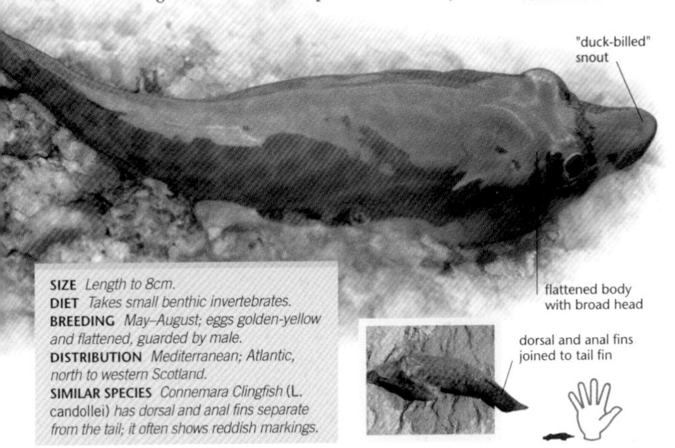

SIZE *Length to 8cm.*
DIET *Takes small benthic invertebrates.*
BREEDING *May–August; eggs golden-yellow and flattened, guarded by male.*
DISTRIBUTION *Mediterranean; Atlantic, north to western Scotland.*
SIMILAR SPECIES *Connemara Clingfish* (L. candollei) *has dorsal and anal fins separate from the tail; it often shows reddish markings.*

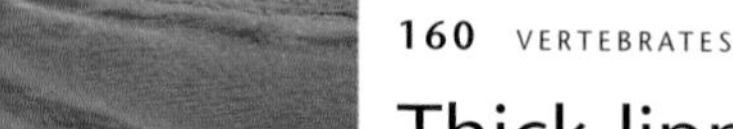

Thick-lipped Grey Mullet

Chelon labrosus (Mugilidae)

FOUND *in shoals in estuaries, lagoons, harbours, and other inshore locations, often muddy or polluted.*

A familiar fish around harbours and marinas, the Thick-lipped Grey Mullet is the most common of a group of very similar grey mullet species. It has a broad, cylindrical, silvery body, with a flattened head and large scales. Its mouth is relatively small, with a swollen upper lip. The lip has a series of wart-like structures, which are used to scrape food off rocks.

SIZE *Length to 60cm, occasionally longer.*
DIET *Algae, invertebrates, and organic detritus.*
BREEDING *December–April.*
DISTRIBUTION *Mediterranean; Atlantic, north to southern Norway.*
SIMILAR SPECIES *Several other grey mullet species; Sea Bass* (Dicentrarchus labrax), *which is often larger, with a bigger mouth and a distinct lateral line.*

Plaice

Pleuronectes platessa (Pleuronectidae)

LIVES *on, or shallowly buried into, sandy sediment in intertidal pools and the sublittoral zone; extends a little into estuarine waters.*

This bottom-dwelling flatfish has its eyes on the upper, right side, the eyes moving into position in the juvenile stage. The upperparts are variable in colour, changing by the expansion and contraction of pigment cells; normally they are brown with orange blotches.

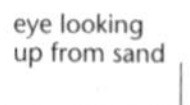

BURIED IN SAND

SIZE *Length to 70cm, occasionally longer.*
DIET *Bottom-dwelling molluscs, crustaceans, worms, and small fish.*
BREEDING *December–April.*
DISTRIBUTION *Western Mediterranean, Atlantic, southern Baltic.*
SIMILAR SPECIES *Flounder* (P. flesus) *is less distinctly spotted; Dab* (Limanda limanda) *has a sandy colour and is often smaller.*

Fifteen-spined Stickleback

Spinachia spinachia (Gasterosteidae)

The only exclusively marine stickleback in Europe, the Fifteen-spined Stickleback has an elongated body with a very long, slender tail stem, and a row of usually fifteen spines on its back, in front of the dorsal fin. The colour of its back is variable according to habitat – greenish among vegetation, brownish on sand and mud. The male takes on a bluish colour at the height of the breeding season. It builds a nest of algal fragments and tends the eggs; the female dies shortly after laying the eggs.

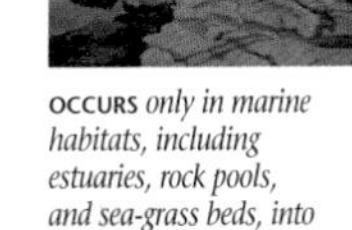

OCCURS *only in marine habitats, including estuaries, rock pools, and sea-grass beds, into the shallow sublittoral.*

tapering snout

long, slender tail stem

SIZE *Length to 19cm.*
DIET *Invertebrates and fish fry.*
BREEDING *April–July, eggs nested in seaweed.*
DISTRIBUTION *Atlantic, north from northern Spain; Baltic.*
SIMILAR SPECIES *Three-spined Stickleback* (Gasterosteus aculeata) *and Nine-spined Stickleback* (Pungitius pungitius), *which are less elongate and have fewer spines.*

Ballan Wrasse

Labrus bergylta (Labridae)

The stout, laterally compressed Ballan Wrasse is very variable in colour – most often it is green or brown with pale spots, but reddish and purple colours often develop, according to its habitat and the breeding stage. Like many related species, the fish starts its life as female, then becomes male as it grows. Unusually among fish, the Ballan Wrasse sleeps on its side.

FAVOURS *rocky, seaweed-covered shores and rock pools, from the lower intertidal to the shallow sublittoral.*

SIZE *Length to 60cm.*
DIET *Molluscs and crustaceans, crushed in the strong jaws; juveniles clean parasites off other fish.*
BREEDING *April–August, eggs laid in a nest of seaweed, held together by mucus threads.*
DISTRIBUTION *Mediterranean (rare), Atlantic.*
SIMILAR SPECIES *Brown Wrasse* (L. merula), *which is a smaller Mediterranean relative.*

Rock Goby

Gobius paganellus (Gobiidae)

INHABITS *weedy rock pools in the intertidal zone, and rocky parts of the shallow sublittoral.*

Gobies are a diverse group of small fish, with a cylindrical body, large fins, and distinctively swollen cheeks. The Rock Goby is one of the more common species – it is fawn-brown in colour with darker mottling and blotches, and has a branched tentacle near the nostril. In the breeding season, the males develop more intense reddish and purple colours.

SIZE *Length to 12cm.*
DIET *Takes small invertebrates.*
BREEDING *January–June; eggs laid in patches under stones.*
DISTRIBUTION *Mediterranean; Atlantic, north to Scotland.*
SIMILAR SPECIES *Sand Goby* (Pomatoschistus minutus) *is paler, with four dark bars; Common Goby* (P. microps) *has dark flank spots.*

Shanny

Lipophrys pholis (Blenniidae)

FAVOURS *exposed rocky shores and rock pools, often stays under seaweed at low tide.*

The commonest member of a large group of rocky shore fish with steeply sloping foreheads and tapering, scale-less, slimy bodies, the Shanny has powerful jaws that are capable of crushing barnacles. The pelvic fins of the Shanny are reduced to two rays, which it uses to "walk" across the surface of rocks.

SIZE *Length to 16cm.*
DIET *Takes invertebrates and green seaweeds.*
BREEDING *April–August; eggs laid under stones or in crevices.*
DISTRIBUTION *Mediterranean (Balearics); Atlantic, north to southern Norway.*
SIMILAR SPECIES *Tompot Blenny* (Parablennius gattorugine) *is larger, with vertical stripes and branched antennae above the eyes.*

Ibiza Wall Lizard

Podarcis pityusensis (Squamata)

FOUND *in areas of bare rock, including coastal cliffs; also widespread in rocky places inland.*

The Ibiza Wall Lizard varies in appearance across different islands and outcrops, and there are 30 recognizable subspecies. On Ibiza, the lizard is normally robust, with coarse, keeled scales, and is brown or green with broken dark stripes on either side of the back. On Formentera, it is slender, and bright green above. Isolated islet populations have strongly developed blue or orange flanks.

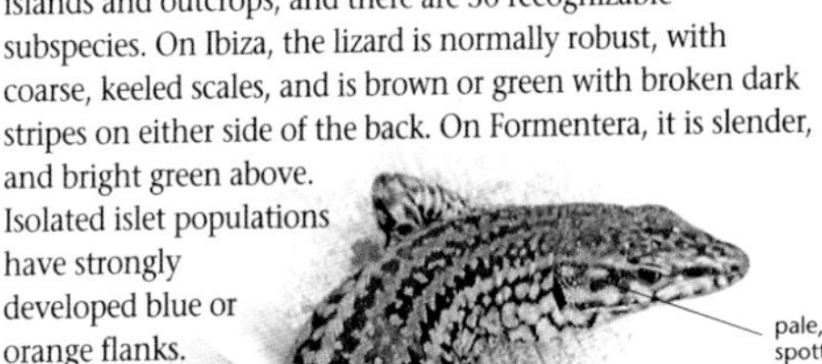

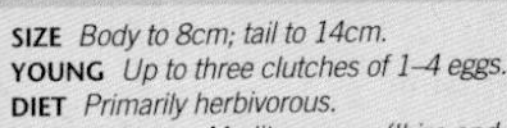

SIZE *Body to 8cm; tail to 14cm.*
YOUNG *Up to three clutches of 1–4 eggs.*
DIET *Primarily herbivorous.*
DISTRIBUTION *Mediterranean (Ibiza and Formentera); introduced locally elsewhere.*
SIMILAR SPECIES *Other wall lizard species are found around the Mediterranean, often in discrete areas. Lilford's Wall Lizard* (P. lilfordi), *on some Balearic islets, is often darker.*

Spiny-footed Lizard

Acanthodactylus erythrurus (Squamata)

OCCUPIES *hot and dry sandy and stony areas with sparse scrub; also found on bare upper beaches.*

The only European representative of a mainly African desert-dwelling group that has toes fringed with scales to enable them to move swiftly over loose sand, the Spiny-footed Lizard is typically found in open, sandy areas, over which it runs fast over long distances. When it stops, it usually adopts a "head-up" alert posture.

distinct groove on top of snout

pale stripes down back

SIZE *Body to 8cm; tail to 15cm.*
YOUNG *One or two clutches of up to eight eggs, hatching in about two months.*
DIET *Insects, especially grasshoppers, other ground-dwelling invertebrates, fruit, and seeds.*
DISTRIBUTION *Western Mediterranean; Atlantic (Portugal and Spain).*
SIMILAR SPECIES *Several other stripy lizards are found in similar habitats in the south.*

JUVENILE

Dice Snake

Nartix tessellata (Reptilia)

FAVOURS *wetland habitats, in lowland, flowing and still waters; tolerant of brackish water by the coast.*

The most aquatic of European snakes, the Dice Snake is named after its dark, square blotches, in rows forming a chequered pattern. It often basks on stones or trees near water but, if disturbed, drops into the water and may remain submerged for several hours. Hibernation takes place in damp holes or crevices; the Dice Snake emerges in April, the male coming out before the female.

pointed snout

grey ground colour

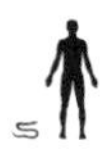

HEAD

SIZE *75–120cm.*
YOUNG *Up to 25 eggs, July–August.*
DIET *Small fish, amphibians, and aquatic invertebrates, ambushed in water.*
DISTRIBUTION *Central and eastern Mediterranean.*
SIMILAR SPECIES N. maura, *which is the western Mediterranean counterpart, and is also tolerant of brackish conditions.*

Balkan Terrapin

Mauremys rivulata (Reptilia)

FOUND *in lowland fresh or brackish water, such as lakes, ponds, and marshes; tolerant of pollution.*

Often seen in brackish, even polluted, coastal wetlands, the Balkan Terrapin has yellow-green stripes on its neck, a feature shared with its more western counterpart, the Spanish Terrapin (*M. leprosa*); this distinguishes both from the more widespread European Pond Terrapin (*Emys orbicularis*). It often basks in the sun on exposed mud banks, but is easily disturbed if approached.

dark eyes

pale stripes on neck

keeled shell

SIZE *Shell to 25cm long.*
YOUNG *Lays 2–3 clutches of up to six eggs.*
DIET *Small fish, amphibians, invertebrates.*
DISTRIBUTION *Mediterranean (Balkans).*
SIMILAR SPECIES *Spanish Terrapin* (M. leprosa), *in similar habitats in Iberia, has pale rather than dark eyes; European Pond Terrapin* (Emys orbicularis), *which lacks neck stripes, and has a shell that is not flared at the back.*

Loggerhead Turtle

Caretta caretta (Reptilia)

Sea turtles are truly oceanic wanderers, rarely approaching land except for breeding. The Loggerhead Turtle is the species most frequently observed in Europe – mostly as juveniles that have drifted to the western seaboard on the Gulf Stream, and as a breeding species on the Canary Islands and (locally) in the Mediterranean. The adult has a very large, elongated oval, red-brown shell, which is made up of horny plates and is paler yellow-green beneath. The only other breeding sea-turtle in Europe, the Green Turtle, is now restricted to a few nesting beaches in the northeast Mediterranean.

LOOK *for tracks of females visiting sandy beaches to breed at night; may be seen by day in shallow coastal waters, or as a tideline stranding.*

NOTE

The Loggerhead Turtle is an endangered species; breeding females may be affected by nocturnal disturbance on the beaches they lay their eggs on. If disorientated by lights from tourist spots, new hatchlings can become vulnerable to predation.

SIZE *Shell up to 1.2m long.*
YOUNG *Up to six clutches of about 200 eggs.*
DIET *Marine crustaceans; plant material.*
DISTRIBUTION *Mediterranean; Atlantic, north to Ireland.*
SIMILAR SPECIES *Three similar species have been recorded in Europe: Kemp's Ridley* (Lepidochelys kempii), *which has a small, broad shell; Green Turtle* (Chelonia mydas), *which is large and oval, breeds sparingly in the Mediterranean, and is common around the Canaries; Hawksbill Turtle* (Eretmochelys imbricata), *which has small, overlapping scales.*

Western Spadefoot

Pelobates cultripes (Amphibia)

FAVOURS *free-draining, sandy soils, including sand dunes and heaths, with access to ponds and pools for breeding.*

Marbled with brown and green, the Western Spadefoot has a modified hind foot, which helps it to burrow into the sandy soil of its preferred habitats. It both hibernates (in winter) and aestivates (during drought) in its burrow. It breeds mainly in temporary pools, which, compared to more permanent waters, are less likely to contain the predators of its tadpoles.

silvery green iris, with vertical pupil

marbled skin

black spade on hind foot

grey-brown ground colour

SIZE *Body 7–10cm.*
YOUNG *Lays up to 7,000 eggs.*
DIET *Insects, snails, and worms.*
DISTRIBUTION *Western Mediterranean, Atlantic (Portugal and Spain).*
SIMILAR SPECIES *Common Spadefoot* (P. fuscus) *in southern Baltic; Eastern Spadefoot* (P. syriacus) *in the Balkans, both of which have pale spades.*

Marsh Frog

Rana ridibunda (Amphibia)

OCCUPIES *usually well-vegetated, permanent water bodies; tolerates brackish conditions by the coast.*

The largest native European frog, this species is also one of the noisiest, with a chorus of rapidly repeated loud croaks rising to a crescendo. Males have grey vocal sacs. Often seen basking on floating leaves, Marsh Frogs have whitish underparts, marbled in olive-green. Highly aquatic, they take refuge in water when disturbed and hibernate underwater.

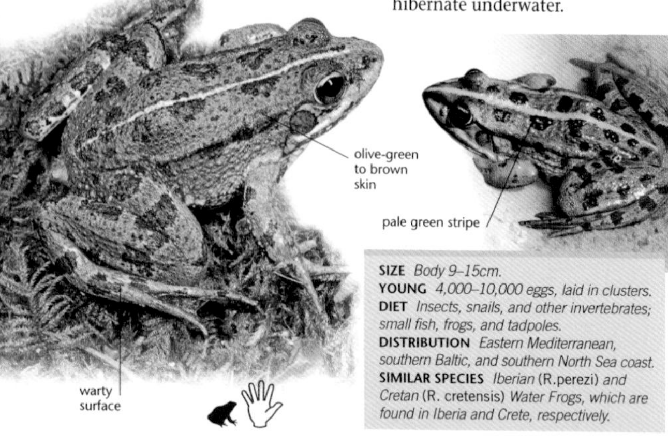

SIZE *Body 9–15cm.*
YOUNG *4,000–10,000 eggs, laid in clusters.*
DIET *Insects, snails, and other invertebrates; small fish, frogs, and tadpoles.*
DISTRIBUTION *Eastern Mediterranean, southern Baltic, and southern North Sea coast.*
SIMILAR SPECIES *Iberian* (R.perezi) *and Cretan* (R. cretensis) *Water Frogs, which are found in Iberia and Crete, respectively.*

Natterjack Toad

Bufo calamita (Amphibia)

A small, but robust toad, the Natterjack Toad is widespread across western Europe, although it has very specific habitat preferences in the north of its range. Here, it frequents sandy areas, where it can burrow for daytime refuge, hibernation, and aestivation, in sites which are often used communally. Its characteristic marking is a yellow stripe down its spine, sometimes appearing very bright (although occasionally absent), and it has parallel parotid glands that are often orange. Its gait is distinctive – the Natterjack scurries like a mouse, rather than walking like most toads.

BREEDS *in pools, often temporary, in dune slacks, upper salt marshes, and other coastal habitats; also widespread inland.*

NOTE

A very noisy species when breeding, males have a vocal sac under the chin, which amplifies their repeated, rolling croak. A chorus of males can be heard at a distance of several kilometres, and may be mistaken for the churring call of a Nightjar.

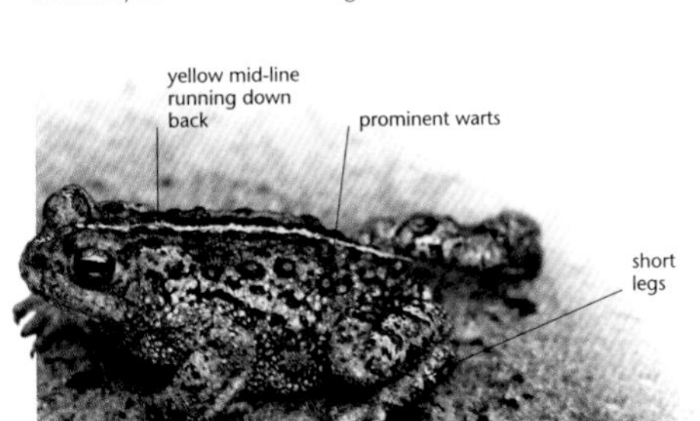

SIZE *Body 6–8cm.*
YOUNG *Lays 2,000–5,000 eggs, hatching rapidly; may develop very quickly in shallow water. Tadpoles are small and black, with bronze spots.*
DIET *Slugs, spiders, insects, and worms.*
DISTRIBUTION *Western Mediterranean; Atlantic, as far north as Denmark; southern Baltic.*
SIMILAR SPECIES *Western (p.166) and other spadefoots, which are often the only toads found in similar coastal locations;* Bufo bufo *and* B.viridis, *which are found in coastal pools, but lack the pale back stripe.*

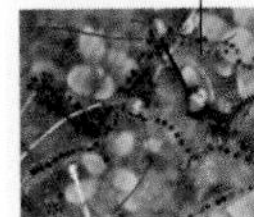

SPAWN

Birds

By virtue of their power of flight, almost all European birds can be found on the coast, whether for breeding, feeding, or on migration. The birds featured here are the ones most strongly associated with coastal habitats in respect of breeding, and those that use the coast primarily as a winter feeding site. The main groups covered are waterfowl, waders, gulls, and their relatives, and a diverse range of seabirds, from cormorants to shearwaters. Almost all birds, apart from those found permanently at sea, require fresh water for drinking. So in salt-dominated coastal environments, you are likely to find birds where there is fresh water.

HERRING GULL

RED-CRESTED POCHARD

GYRFALCON

OYSTERCATCHER

Gannet

Morus bassanus (Sulidae)

The largest of all European seabirds, the adult Gannet is typically seen offshore as a brilliant white bird with black wingtips, flying steadily, singly or in straggling lines, or circling and diving for fish with spectacular plunges. Juvenile birds, by contrast, are very dark with white specks, gradually becoming whiter over about five years, but their shape is always distinctive. At close range, the breeding adult's yellow-buff head is visible, as well as the dark markings on its face and fine lines on its big, sharp bill.

BREEDS *in dense, noisy colonies on rocky islands and cliffs. Feeds at sea, many moving south for the winter.*

VOICE *Rhythmic, throaty chorus of grunts and croaks at nest; otherwise silent, except cackling from feeding groups at sea.*
FEEDING *Mackerel, Pollack, and other fish, caught by plunge-diving; also scavenges from fishing boats.*
DISTRIBUTION *Breeds: Atlantic, northwards from Brittany. Winters: Atlantic and western Mediterranean.*
SIMILAR SPECIES *Adults are unmistakable, but young birds resemble the immature Great Black-backed Gull* (Larus marinus), *which is almost as large, but has less pointed wings and head.*

NOTE

A feeding frenzy of plunging Gannets indicates the presence of a shoal of fish: a good place to look out for other fish-eaters, such as dolphins.

Herring Gull

Larus aregentatus (Laridae)

The big, noisy Herring Gull is mainly a bird of sea cliffs in summer, but roams over all kinds of shores and far inland in winter, when its white head and neck are streaked brownish. Paler than the Yellow-legged Gull (below), with pink legs, it has fierce-looking pale eyes.

FEEDS *on beaches, estuaries, reservoirs, and refuse tips. Breeds on cliffs, islands, and rooftops.*

VOICE *Squealing notes, yelps, and barks.*
FEEDING *Takes fish, invertebrates, carrion, offal, and scraps from ground or water.*
DISTRIBUTION *Atlantic, Baltic.*
SIMILAR SPECIES *Yellow-legged Gull (below); Common Gull* (L. canus), *which is smaller with yellow-green bill and legs; Great Black-backed Gull* (L. marinus), *which is larger with black wings and back.*

Yellow-legged Gull

Larus (cachinnans) michahellis (Laridae)

Formerly regarded as a southern subspecies of the Herring Gull (above), this bird has a darker back, less white at the wingtips, and yellow instead of dull pinkish legs. Its head remains white in winter, but is streaked pale grey in autumn. In some places, the two species breed side-by-side, without hybridizing.

BREEDS *on rocky islands and offshore stacks. Frequently scavenges around docks and towns.*

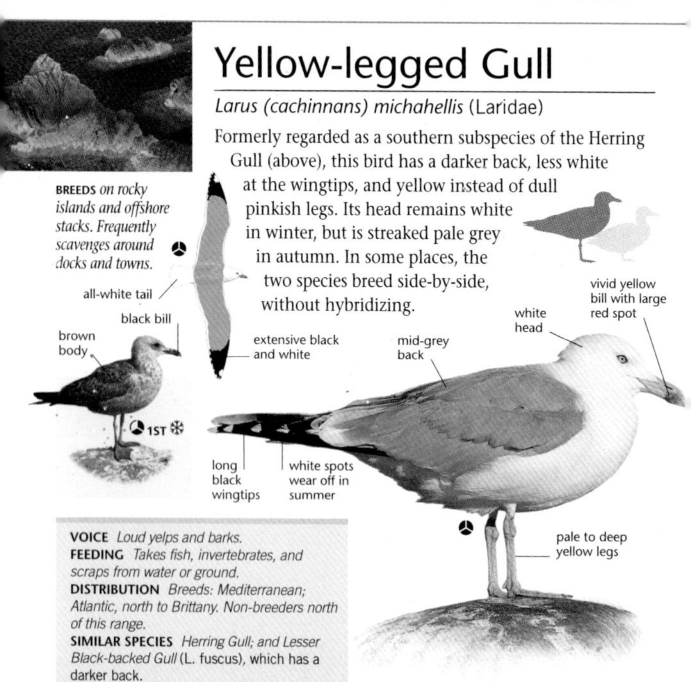

VOICE *Loud yelps and barks.*
FEEDING *Takes fish, invertebrates, and scraps from water or ground.*
DISTRIBUTION *Breeds: Mediterranean; Atlantic, north to Brittany. Non-breeders north of this range.*
SIMILAR SPECIES *Herring Gull; and Lesser Black-backed Gull* (L. fuscus), which has a darker back.

Kittiwake

Rissa tridactyla (Laridae)

One of the most maritime of gulls, the Kittiwake comes to land only to breed in noisy colonies on sheer cliffs. Its very white head and black wingtips are distinctive in flight, and its call in flight is unmistakable. In winter, the adult has a grey nape and a dark ear patch.

BREEDS *on rocky cliffs, and ledges on buildings near the sea; winters at sea, well away from the coast.*

VOICE *Ringing, repeated* kitti-i-waake.
FEEDING *Fish from sea, offal from trawlers.*
DISTRIBUTION *Atlantic, breeds mainly from Brittany northwards with outliers in Spain.*
SIMILAR SPECIES *Immature Little Gull* (L. minutus), *which is similar to immature Kittiwake, but is smaller with a more buoyant flight; adults are distinguished by a black hood (in summer) and dark underwings.*

Great Skua

Stercorarius skua (Stercorariidae)

WINTERS *at sea, usually, well offshore. Breeds on cliff-top moorland and rocky slopes.*

The largest, heaviest, boldest, and most predatory of the skuas, the Great Skua is always dark brown with pale buff streaks and big white wing patches. Able to steal from a Gannet and kill a Kittiwake, its success in recent years has caused problems for other seabirds. In the northern Atlantic islands, it is known as "Bonxie", which means an untidy old woman.

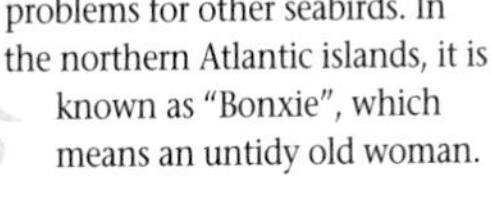

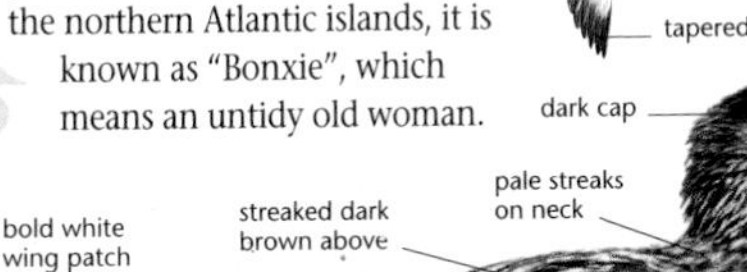

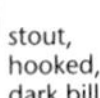

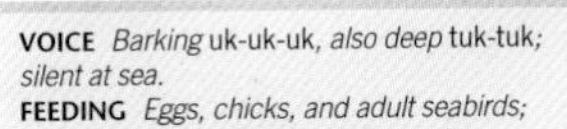

VOICE *Barking* uk-uk-uk, *also deep* tuk-tuk; *silent at sea.*
FEEDING *Eggs, chicks, and adult seabirds; food stolen from other seabirds; carrion.*
DISTRIBUTION *Breeds: Atlantic. Winters: Atlantic and western Mediterranean.*
SIMILAR SPECIES *Other skuas, especially juveniles and dark-phase adults, which are smaller, but still have a white wing-flash.*

Audouin's Gull

Larus audouinii (Laridae)

An extremely rare gull, this bird breeds in limited sites around the Mediterranean. However, its numbers have increased in recent years, and its wintering range has expanded southwards. It can be spotted in favoured haunts, such as harbours in the Balearics, where it can be easily identified by its dark bill and eyes.

NESTS *on small, rocky islands and islets, also locally on cliffs and salt marsh; more widespread around the coast outside the breeding season.*

VOICE *Series of nasal, raucous squawks, often in an extended sequence.*
FEEDING *Fish, snatched from sea in flight; sometimes plunge-dives; rarely scavenges.*
DISTRIBUTION *Mediterranean.*
SIMILAR SPECIES *Yellow-legged Gull (p.170) and in winter Lesser Black-backed Gull* (L. fuscus)*, which are larger and darker, with yellow bill.*

Mediterranean Gull

Larus melanocephalus (Laridae)

Although more common in southeast Europe, this beautiful gull has spread west, far beyond its original range. Bigger than the similar Black-headed Gull (p.173), with a heavier bill, it has white wingtips instead of black, and a jet-black rather than dark brown hood in summer.

BREEDS *on shallow lagoons and coastal marshes. Winters on estuaries, beaches, and rarely on lakes.*

VOICE *Nasal, far-carrying, rising, and falling* eeu-err eeu-err.
FEEDING *Forages for fish, worms, aquatic invertebrates, and offal on beaches and land.*
DISTRIBUTION *Mediterranean; Atlantic, north to Denmark; few breeding colonies.*
SIMILAR SPECIES *Black-headed Gull (p.173); immature resembles immature Common Gull* (L. canus).

Black-headed Gull

Larus ridibundus (Laridae)

Common and familiar, this is a small, agile, very white-looking gull. It is never truly "black-headed", because in breeding plumage its hood is dark chocolate brown and does not extend to the back of its head. In other plumages, it has a pale head with a dark ear spot. Its dark underwing gives a flickering effect in flight. It has always been a frequent bird inland, but numbers have increased still further in response to abundant food provided by refuse tips and safe roosting sites on reservoirs and flooded pits.

BREEDS *on high-level salt marshes and around coastal lagoons, also on upland moorland; widespread inland and around the coast in winter.*

NOTE

Look out for a striking white wedge-shaped marking along the leading edge of each outer wing; no other common gull has this feature.

VOICE *Harsh squealing, laughing and chattering calls, especially around breeding colonies.*
FEEDING *Worms, fish, insects, seeds, and scraps taken from the sea, the ground, or in the air.*
DISTRIBUTION *Breeds: western Mediterranean, Atlantic, Baltic. Winters: throughout the region.*
SIMILAR SPECIES *Mediterranean Gull (p.172); Little Gull* (L. minutus), *which is smaller, the adult having blackish underwings; Slender-billed Gull* (L. genei), *which lacks head markings, and is found in the Mediterranean.*

Fulmar

Fulmarus glacialis (Procellariidae)

FEEDS *at sea; breeds on rocky cliffs, remote grassy cliff slopes, and in hollows excavated in soft, sandy cliffs.*

Although it has gull-like plumage, the Fulmar is a tube-nosed petrel, more closely related to the albatrosses. It has a distinctive thick neck and black eye patch and holds its wings straight when gliding, unlike a gull. The tube nose contains salt excretion glands, an essential adaptation for a bird that drinks only salt water and eats only seafood. Fulmars are often seen soaring on updraughts around coastal cliffs, although they spend much of their time at sea. Originally an Arctic bird, the Fulmar's distribution has extended over the past century, due in part to its habit of following fishing boats back to port in search of offal. It nests on cliff or earth ledges, or rarely, on buildings.

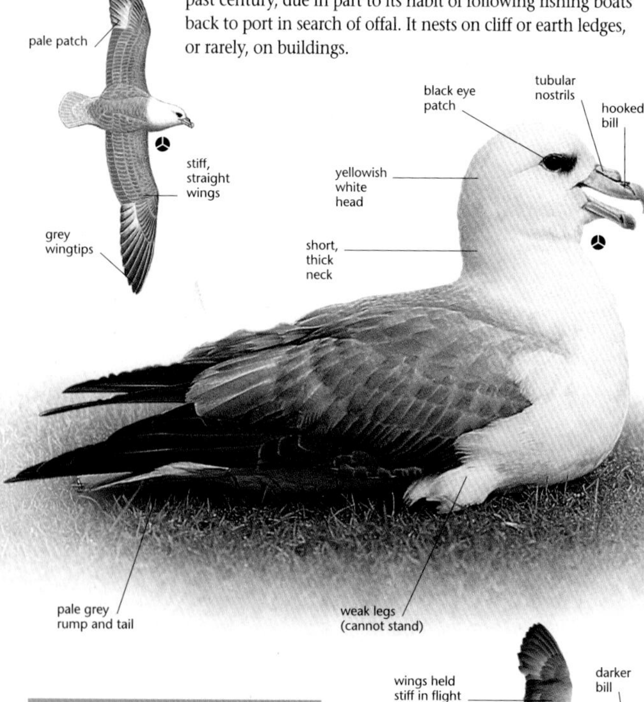

NOTE

Fed on rich fish oils, immature Fulmars gain weight rapidly. They remain in the nest for several weeks after their parents depart to offshore wintering grounds. These oils are also an effective defence against predators, being regurgitated when a bird is threatened.

VOICE *Loud, harsh, throaty cackling.*
FEEDING *Takes mostly fish offal from trawlers; small fish, jellyfish, squid, and other marine animals.*
DISTRIBUTION *Atlantic, north from Brittany; further south, but mostly offshore, in winter.*
SIMILAR SPECIES *Herring Gull (p.170), which lacks tubular nostrils and the stiff-winged flight; Cory's Shearwater (p.175), which is darker above and more elegant, with longer wings, preferring a distinctive "looping" flight.*

Cory's Shearwater

Calonectris diomedea (Procellariidae)

A big, brown-backed, long-winged ocean bird, Cory's Shearwater has a distinctive yellowish bill and a dark smudge around its eye. It often holds its long wings gently arched, and when extended, they show a faint, dark "W" mark. It soars high in the air, then swoops down in lazy, rolling glides, swerving slowly in long, banking arcs.

long, tapered, slightly rounded wings

LIVES *mainly well out at sea, but sometimes occurs close inshore off headlands and islands. Breeds on many of the Mediterranean islands.*

VOICE *Loud, varied wailing sounds near breeding sites, at night; mostly silent at sea.*
FEEDING *Takes fish, squid, shrimps, and jellyfish in shallow dives from the surface of sea, and waste from fishing vessels.*
DISTRIBUTION *Mediterranean; also in Atlantic, north to Britain (non-breeding season).*
SIMILAR SPECIES *Immature Herring Gull* *(p.170)**, and Fulmar* *(p.174)**.*

Manx Shearwater

Puffinus puffinus (Procellariidae)

Ungainly on land, shuffling along on its weak legs, aided by its bill and wings, this is a swift, elegant bird in the air. It is usually seen flying low over the sea with rapid, stiff-winged flaps between long glides, flashing alternately black and white as it banks from side to side.

stiff wings

FEEDS *over open sea, and breeds in colonies on islands and remote headlands. Widespread when migrating off coasts in autumn.*

white flank each side of rump

black cap

VOICE *Wailing and chortling sounds around breeding colonies at night.*
FEEDING *Fish and small squid, taken by surface- or plunge-diving.*
DISTRIBUTION *Breeds: Atlantic, from Brittany to Faeroes.*
SIMILAR SPECIES *Balearic Shearwater* (P. mauretanicus) *is larger and browner; Yelkouan Shearwater* (P. yelkouan) *(Mediterranean).*

Sandwich Tern

Sterna sandvicensis (Sternidae)

BREEDS *on sand dunes, shingle beaches, and islands. Winters in coastal waters of Africa.*

The Sandwich Tern is a large, active, noisy bird with a spiky black crest, a long, sharp bill, and long, angular wings that it often holds away from its body, slightly drooped. It looks very white in the air, diving for fish from high up and hitting the water with a loud smack.

VOICE *Loud, rasping screech.*
FEEDING *Catches fish, especially sand eels, by plunge-diving from the air.*
DISTRIBUTION *Throughout, north to Scotland and southern Baltic. Winters: very locally in the Mediterranean.*
SIMILAR SPECIES *Gull-billed Tern* (S. nilotica), *which has a short, stout, black bill, and is largely Mediterranean.*

Caspian Tern

Sterna caspia (Sternidae)

BREEDS *on coasts and low islands, mostly in Baltic; rare migrant on other coasts.*

The Caspian Tern is enormous compared to other terns, with a massive red bill. Yet it is a well-proportioned bird, and if there are no other terns for comparison its bulk may not be obvious. Now scarce, it can be seen flying steadily over water, head angled down, looking for fish.

VOICE *Very loud, deep, rasping cry.*
FEEDING *Plunge-dives for fish; may fly long distances from breeding colony to feed.*
DISTRIBUTION *Breeds in Baltic, very locally in eastern Mediterranean; a few winter in the Mediterranean and the Algarve.*
SIMILAR SPECIES *Lesser Crested Tern* (S. bengalensis), *which has an orange bill, and is found on the south Mediterranean coast.*

Little Tern

Sterna albifrons (Sternidae)

The quick, nervous, and tiny Little Tern is usually easy to identify by its size alone, but it is also paler than other terns, with a white forehead all year round. Its wings have a black streak at the tip on the upper side, and the adult has a distinctive yellow dagger bill with a black tip. Although a widespread summer visitor, the Little Tern is becoming scarce on many coasts because it often breeds on popular leisure beaches; many colonies survive only because they are fenced off and protected. Rising sea levels may also be a threat, since extra-high tides often destroy its eggs and chicks.

WINTERS *along the coasts of Africa. Breeds on narrow sand and shingle coastal beaches; also inland in Spain, Portugal, and eastern Europe.*

black streak at wingtips

short, white, forked tail

pure white underside

long, narrow wings

white forehead

black stripe through eye

black cap

black nape

sharp yellow bill with tiny dark tip

pale grey back

black wedge at wingtips

orange to yellow legs

streaky crown

dark chevrons on back

NOTE

Squeezed into dense colonies by human beach activities, Little Tern eggs and chicks face many dangers. They are often heavily preyed upon, by both mammals (such as foxes, hedgehogs, and Badgers) and birds, such as crows and kestrels. Add the risks from high tides and trampling, it is not surprising that the Little Tern is declining over much of its range.

VOICE *Sharp, high, rapid chattering.*
FEEDING *Plunge-dives for small fish, after a rapid hover, often in very shallow water and even in breaking waves.*
DISTRIBUTION *Breeds: throughout, except northern Baltic. Winters: off west Africa.*
SIMILAR SPECIES *Three larger terns – Roseate Tern* (S. dougallii), *which has a black bill with a red base; Arctic Tern* (S. paradisaea), *which has an all-red bill; and Common Tern* (S. hirundo), *which has a red bill with a black tip.*

Guillemot

Uria aalge (Alcidae)

BREEDS *in colonies on narrow ledges on sea-cliffs and flat-topped stacks. Winters at sea, well offshore.*

A slim, long-bodied auk with a slender, pointed bill, the Guillemot is one of the commonest breeding seabirds at cliff colonies. It often nests alongside the more thickset Razorbill (*Alca torda*), which has a deeper bill. Guillemots can often be seen flying low and fast off headlands, or swimming in large groups below the cliffs. As with all auks, they are very vulnerable to marine pollution, and the more southerly populations have declined very severely, partly as a result of oil spills.

NOTE

Northern subspecies have a much darker brown, almost black, back than southern birds. The bridled form, which has a white eye-ring in summer, is common to all subspecies, but its frequency increases northwards, with up to 25 per cent occurrence in Norway.

VOICE *Loud, growling chorus at breeding colonies.*
FEEDING *Dives from surface for fish deep underwater, propelled by wings.*
DISTRIBUTION *Atlantic and Baltic. Winters: away from breeding colonies, but mostly offshore.*
SIMILAR SPECIES *Black Guillemot* (Cepphus grylle), *which is chocolate-brown in summer, with white wing patches and red feet; Brunnich's Guillemot* (U. lomvia), *which has a shorter, stouter bill, and is found in the Arctic; and Razorbill* (Alca torda), *which has a deeper bill.*

Puffin

Fratercula arctica (Alcidae)

Few seabirds are more instantly recognizable than the Puffin in summer, with its clown-like eye and huge, flamboyantly coloured bill. Even at a distance it is usually distinctive, bobbing on the water or whirring through the air like a clockwork toy. In winter, it is less striking, as the colourful eye ornaments and horny plates at the edges of its bill fall away; its face is also darker, although not as dark as that of a juvenile bird. In summer, Puffins are often to be seen bringing food to their nesting burrows on northern and western coasts, but Puffins in winter plumage are generally rare close inshore.

BREEDS *on coastal clifftops, mainly on islands, feeding in nearby waters. Winters well out to sea.*

plain black wings

black sides to rump

dusky underwing

dark eyes, bluish scale above

triangular bill, striped bluish, orange, yellow, and red

grey-white facial disc

black upperparts and neck

white below

dusky grey face

smaller, duller bill

NOTE

A breeding Puffin is hard to mistake for any other bird, but in winter, Puffins usually feed well offshore and may be confused with other auks at long range. One distinctive feature is the grey facial disc; a Puffin also has a dumpier, more front-heavy body than a Guillemot (p.178) or a Razorbill (Alca torda), *and it has all-dark wings with no white trailing edge.*

VOICE *Loud, cackling growl at nest; otherwise generally silent.*
FEEDING *Dives from water surface to catch small fish such as sand eels; also takes small squid, crustaceans, and marine worms.*
DISTRIBUTION *Breeds: Atlantic, north from Brittany. Winters: Atlantic and western Mediterranean.*
SIMILAR SPECIES *Little Auk* (Alle alle), *which winters in the north Atlantic, is smaller, and lacks the ornate beak; Guillemot (p.178) and Razorbill* (Alca torda), *which show white sides to the rump in flight and are larger.*

Shag

Phalacrocorax aristotelis (Phalacrocoracidae)

FEEDS *off rocky coasts and islands. Breeds on coastal cliffs.*

A large, long-bodied diving bird, the adult Shag is black overall with an oily green gloss. It also has a yellow patch at the base of its bill. In summer, both sexes sport a short, curly crest. Winter birds and immatures are less distinctive and may be confused with Cormorants (below), although all have a more slender form and a steeply sloping forehead.

VOICE *Grunts, hisses, and coarse, frenzied rattling at nest; silent when feeding at sea.*
FEEDING *Fish, mainly sand eels and herrings, caught underwater after a surface dive.*
DISTRIBUTION *Mediterranean, Atlantic.*
SIMILAR SPECIES P. a. desmarestii *is the Mediterranean subspecies; Cormorant (below); Pygmy Cormorant* (P. pygmaeus) *is smaller and restricted to southeast Europe.*

Cormorant

Phalacrocorax carbo (Phalacrocoracidae)

BREEDS *on cliffs and, inland, in trees and reedbeds, and on the ground; feeds in estuaries, bays, harbours, and inland waters.*

Bigger and bulkier than a Shag (above), the Cormorant has a flatter forehead and no crest. In summer, it develops white streaks on the head and white thigh patches, especially prominent in the continental subspecies *P. c. sinensis*. Cormorants swim with their backs almost awash, and frequently perch with wings half open, to dry after diving. They fly with long, strong glides, often high up, unlike Shags.

VOICE *Growling and cackling at nests and communal roosts, but otherwise a quiet bird.*
FEEDING *Dives for fish, especially bottom-living flatfish and eels, in long underwater dive from surface, propelling itself with its feet; brings larger fish to surface before swallowing them.*
DISTRIBUTION *Throughout the region.*
SIMILAR SPECIES *Shag (above).*

Long-tailed Duck

Clangula hyemalis (Anatidae)

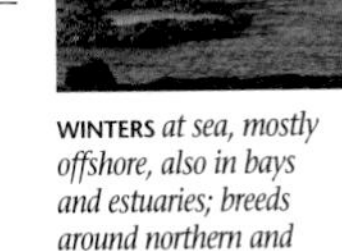

WINTERS *at sea, mostly offshore, also in bays and estuaries; breeds around northern and Arctic inland waters.*

Exclusively marine outside the breeding season, Long-tailed Ducks are unique in that both males and females have distinct summer and winter plumages. Winter males are particularly striking, their largely white plumage set off by blackish markings and a pair of long, dark central tail feathers. Long-tailed Ducks often fly low over the water, splash down, and take off again. They feed in flocks, often associating with Common Scoters (p.183).

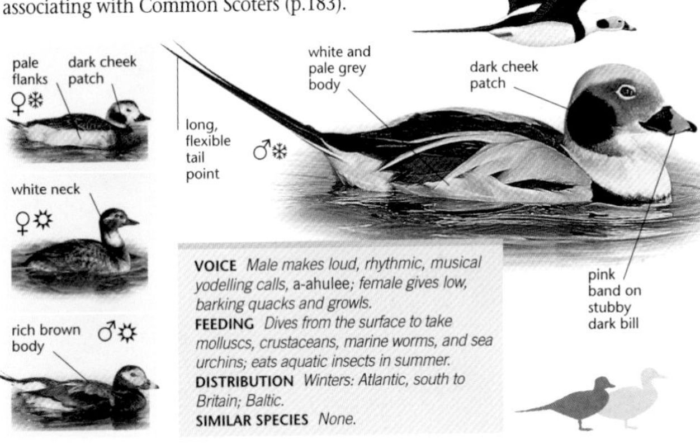

VOICE *Male makes loud, rhythmic, musical yodelling calls,* a-ahulee; *female gives low, barking quacks and growls.*
FEEDING *Dives from the surface to take molluscs, crustaceans, marine worms, and sea urchins; eats aquatic insects in summer.*
DISTRIBUTION *Winters: Atlantic, south to Britain; Baltic.*
SIMILAR SPECIES *None.*

Red-throated Diver

Gavia stellata (Gaviidae)

BREEDS *by remote moorland pools and lakes; feeds at sea; winters on coasts and estuaries.*

A long-bodied, skilled swimmer and diver, the Red-throated Diver usually has a low profile in the water, and is rarely seen on land except at the nest. The smallest diver, it is distinguished by its slimmer bill, held angled upwards. In summer, it is often seen commuting between its nest and the sea, flying high with outstretched head and legs, calling loudly.

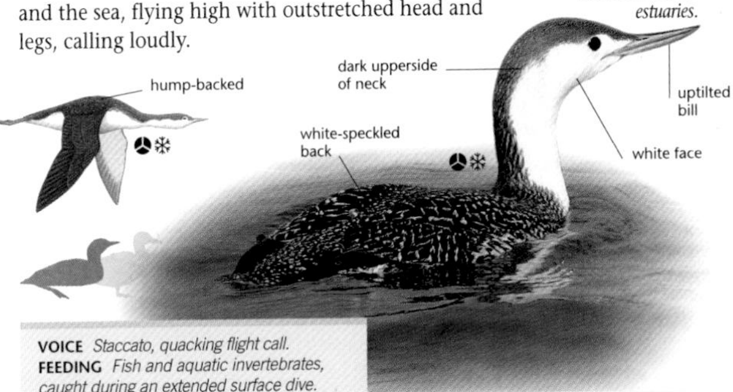

VOICE *Staccato, quacking flight call.*
FEEDING *Fish and aquatic invertebrates, caught during an extended surface dive.*
DISTRIBUTION *Breeds: Atlantic, Baltic. Winters: Atlantic, locally in Mediterranean.*
SIMILAR SPECIES *Black-throated* (G. arctica) *and Great Northern* (G. immer) *Divers, Great Crested Grebe* (Podiceps cristatus), *and Red-breasted Merganser* (Mergus serrator).

pale dusky face

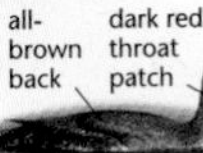

Eider

Somateria mollissima (Anatidae)

BREEDS *on low-lying northern coasts and islands with rocky shores and weedy bays. Winters at sea, often in sandy bays and over mussel beds.*

A big, bulky, entirely marine duck with a characteristic wedge-shaped head, the Eider is usually easy to identify. A winter male is boldly pied black and white, with green patches on its head and a pink flush on its breast. Females have brown plumage with close dark bars that provide superb camouflage on the nest. Highly sociable, Eiders often form large rafts offshore, but they are equally familiar around coastal rocks.

VOICE *Male gives sensuous, cooing* aa-ahooh*; females respond with deep growls and a mechanical* kok-kok-kok.
FEEDING *Molluscs, especially mussels, crustaceans, and other invertebrates gathered during a surface dive.*
DISTRIBUTION *Breeds: Atlantic, north from France, recently in northern Adriatic; Baltic. More widespread in south of range in winter.*
SIMILAR SPECIES *Common Scoter (p.183);* King Eider (S. spectabilis) *and Steller's Eider* (Polysticta stelleri) *of the Arctic, which winter off Norway and locally in the Baltic.*

NOTE

Eiders use neck and head muscles to pull mussels from rocks. These are swallowed whole, crushed in the gizzard, and the shell remnants are produced as a pellet.

Common Scoter

Melanitta nigra (Anatidae)

A dark, large-bodied, slim-necked sea duck with a pointed tail, the Common Scoter gathers in large sociable groups out at sea, swimming buoyantly or flying low over the waves. The male is the only totally black duck; the female is browner with a pale face, but has no truly white body plumage.

WINTERS *at sea around coasts, often well offshore. Breeds on shores of moorland lakes and pools. Occurs in large flocks at regular sites.*

pale tip to underwings
♂
slim neck
black body with duller or paler wings
♂❄
round head
pointed bill with yellow patch
thin neck
long, pointed tail, often raised
dark cap
♀ dark brown body
grey face

VOICE *Male has musical, piping whistle; female makes deep growls.*
FEEDING *Dives from surface to find shellfish, crustaceans, and worms.*
DISTRIBUTION *Winters: Atlantic; locally in the Mediterranean; on passage in the Baltic.*
SIMILAR SPECIES *Velvet Scoter* (M. fusca), *which has white wing patches; the male has a white eye-spot and yellow sides to the bill.*

Scaup

Aythya marila (Anatidae)

With a rounder crown than the Tufted Duck (*A. fuligula*), and no tuft, the Scaup is also more marine in its habits. The black head of a breeding male has a green gloss in good light, while the back is a pale, marbled grey rather than black. Winter flocks of Scaup favour sheltered waters, where the white flanks of the males show up well against the dark sea.

WINTERS *in quiet coastal waters such as estuaries. Breeds by lakes and pools.*

VOICE *Male produces low whistles in display, otherwise mostly silent; female has deep growl.*
FEEDING *Dives for molluscs, crustaceans, other invertebrates; aquatic plants.*
DISTRIBUTION *Breeds: Baltic, inland in Norway. Winters: Atlantic, south to France.*
SIMILAR SPECIES *Tufted duck* (A. fuligula), *which has a black-backed male, while the female has little or no white face patch.*

Wigeon

Anas penelope (Anatidae)

The white wing patches of the male Wigeon and white belly of the female Wigeon are prominent even at a distance. The Wigeon feeds on grasses and other plants like a miniature goose. The flocks take to the air at the first sign of danger, wheeling about as the males let out their far-carrying calls.

WINTERS *in flocks on estuaries and salt marshes, as well as inland waters; breeds inland, around northern pools and lakes.*

VOICE *Male has explosive, musical whistling whee-oo; female gives abrupt, harsh growls.*
FEEDING *Eel-grasses and green seaweeds; dabbles and upends in shallow water for seeds.*
DISTRIBUTION *Winters: Atlantic, north to southern Norway; Mediterranean.*
SIMILAR SPECIES *Teal* (A. crecca), *which is smaller, but often occurs with Wigeon, females lack the clear white belly.*

Red-crested Pochard

Netta rufina (Anatidae)

Unevenly distributed across Europe, the Red-crested Pochard is particularly found around the Mediterranean. The male, with its rust-orange head and red bill, is unmistakable; the female has a pale face patch, similar to that of a female Common Scoter (p.183), though their ranges and habitats barely overlap.

FAVOURS *coastal lagoons and deltas, as well as shallow inland waters; prefers reed-fringed sites.*

VOICE *Various sneezing and barking sounds.*
FEEDING *Aquatic vegetation and leaves, obtained by upending, dabbling, and diving.*
DISTRIBUTION *Breeds: locally around the Mediterranean, and Atlantic, north to Denmark; more widespread in winter, but only in the south of its range.*
SIMILAR SPECIES *Female Common Scoter (p.183); Pochard* (Aythya ferina).

Shelduck

Tadorna tadorna (Anatidae)

Looking black and white at a distance, but revealing a bright red bill and rich chestnut patches at close range, the Shelduck is a handsome, erect, rather goose-like duck in breeding plumage. Usually seen in pairs or small, loose flocks, its bright white plumage is easily visible at a distance against the dark mud of an estuary at low tide. Its pattern is also striking during its strong, fast, but rather heavy flight. Family groups gather together in late summer, when most of the adults fly to the North Sea to moult.

FEEDS *on sandy or muddy shores, especially in sheltered estuaries; breeds in grazing marsh and inland wetlands.*

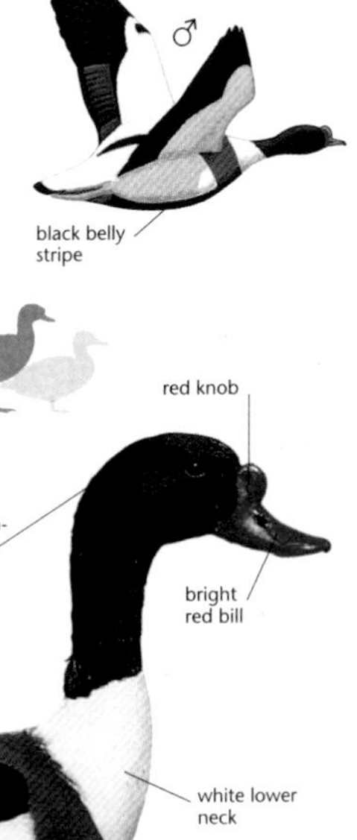

VOICE *Goose-like honk and growl; whistling notes from the male and rhythmic quacking from the female in spring.*
FEEDING *Small snails and other invertebrates, collected from mud in a side-to-side sweeping action; grazes and upends for algae and other plant material.*
DISTRIBUTION *Breeds: Atlantic, north from France, southern Baltic; more widespread in winter, including locally around the Mediterranean.*
SIMILAR SPECIES *Ruddy Shelduck* (T. ferruginea)*, which is similarly bulky, but mostly orange-brown and occurs in the eastern Mediterranean.*

NOTE

After hatching, the downy chicks are led by the female to a nursery area, where they may mix with several other broods, forming large crèches.

Brent Goose

Branta bernicla (Anatidae)

WINTERS *on muddy estuaries, salt marshes, and nearby grazing and arable land; breeds on Arctic tundra.*

A small, very dark goose, the Brent Goose occurs as two main subspecies. The dark-bellied subspecies breeds in Siberia, while the pale-bellied one breeds from Spitzbergen to Arctic Canada. The black-bellied North American subspecies, the Black Brant (*B. b. nigricans*), is a rare vagrant. All subspecies have black heads, and white neck markings, except as juveniles. Highly gregarious birds, they feed in flocks, grazing coastal fields or upending like ducks in shallow water, when the white stern is particularly noticeable.

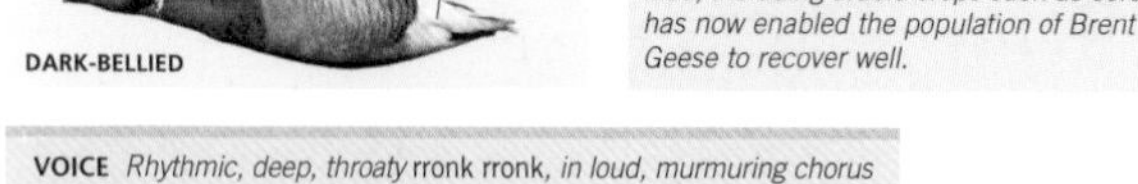

NOTE

Since the primary food of Brent Geese is Eel-grass (p.54), these birds suffered a sharp decline in the early 20th century, when their food source was almost destroyed by a wasting disease. The adoption of alternative food, including arable crops such as cereals, has now enabled the population of Brent Geese to recover well.

VOICE *Rhythmic, deep, throaty* rronk rronk, *in loud, murmuring chorus from flocks.*
FEEDING *Eats eel-grass and algae on coastal mudflats; increasingly cereals and grass.*
DISTRIBUTION *Dark-bellied subspecies winters from the Netherlands and eastern Britain to western France; migrates through the Baltic. Light-bellied subspecies winters in Denmark, northeast Britain and Ireland, with a few elsewhere.*
SIMILAR SPECIES *Barnacle Goose (p.187).*

Barnacle Goose

Branta leucopsis (Anatidae)

Although clearly related to the Canada Goose (*B. canadensis*), this highly social bird is easily identified by its creamy white face, black breast, and beautifully barred back. Juveniles are duller, with less barring. Large flocks winter on the same sites every year, often grazing at night.

WINTERS *on pastures and salt marshes, on traditional sites. Breeds on northern coasts and Arctic tundra.*

VOICE *Harsh, short bark, creating chattering, yapping chorus from flocks.*
FEEDING *Large flocks graze on grass, clover, and similar vegetation.*
DISTRIBUTION *Breeds locally around the Baltic; winters on the Atlantic shores, from Ireland to Denmark.*
SIMILAR SPECIES *Brent Goose (p.186); Canada Goose, which is larger.*

Pink-footed Goose

Anser brachyrhynchus (Anatidae)

This round-headed, short-billed goose has a shorter neck than other geese, and a strong contrast between its very dark head and pale breast – features that are often obvious in flight. It occurs in tens of thousands at favoured sites, feeding in dense flocks by day and making spectacular mass flights to its roosts in the evening.

FEEDS *on marshes, pasture, and arable land near coast. Roosts on lakes, estuaries, and low-lying islands.*

VOICE *Resonant* ahng-unk *and frequent, higher* wink-wink.
FEEDING *Eats grass, waste grain, sugar beet tops, carrots, and potatoes; feeding in flocks.*
DISTRIBUTION *Winters: Atlantic, from Britain to Denmark.*
SIMILAR SPECIES *Other "grey" geese such as White-fronted* (A. albifrons), *Greylag* (A. anser), *and Bean* (A. fabalis) *Geese.*

Oystercatcher

Haematopus ostralegus (Haematopodidae)

ROOSTS *in often huge flocks. Breeds on sandy, muddy, and rocky shores; some inland on grass or river shingle.*

With its dazzling black and white plumage and stout, carrot-coloured, blade-like bill, the Oystercatcher is one of Europe's most unmistakable, and common, waders. A noisy bird, its loud, piercing calls are equally distinctive. The powerful bill is adapted for opening cockles, mussels, and other bivalve molluscs; some birds specialize in hammering the shells until they break, others have a stabbing technique, aiming for the muscle which holds the shells together. Individual feeding techniques are generally learned from the parents.

red eyes with orange eye-ring

big, bright orange-red bill

portly, black and white body

NOTE

A breeding bird on Atlantic and Baltic coasts, the eggs of the Oystercatcher are well camouflaged against the backdrop of stones, where the nest is located. This protects the eggs from predators.

white "V"

broad white wingbar

short, sturdy, pale pink legs; duller in juvenile

dark tip to bill

bill with dark tip

white collar

VOICE *Loud, strident* klip, kleep, *or* kleep-a-kleep; *shrill chorus from big flocks.*

FEEDING *Probes for molluscs and marine worms; prises bivalve molluscs from rocks and seaweed; eats earthworms inland.*

DISTRIBUTION *Breeds: throughout, but only local south of Brittany. Winters: more widespread areas, south from southern Norway.*

SIMILAR SPECIES *Lapwing* (Vanellus vanellus) *is another large wader, which is dark above and white below, but has a short, dark bill; Black-tailed Godwit (p.195), which has similar wing and tail markings in flight.*

Turnstone

Arenaria interpres (Scolopacidae)

Most waders like to feed on soft mud or sand, but the stocky, short-billed Turnstone favours areas of stones, weed, or other debris that it can flick through in search of small animal food. Noisy, active, and often tame, it is colourful in summer but very dark above in winter, with a piebald look in flight.

FEEDS *on sea coasts, especially rocky shores and gravelly tidelines. Breeds on rocky northern coasts.*

VOICE *Fast, hard, abrupt, staccato calls,* tukatukatuk, teuk, tchik.
FEEDING *Stirs up and turns seaweed, shells, and stones to find invertebrates. Highly adaptable in its feeding habit.*
DISTRIBUTION *Breeds: Atlantic (Norway) and Baltic. Winters: Atlantic, south from Britain, locally around the Mediterranean.*
SIMILAR SPECIES *None.*

Purple Sandpiper

Calidris maritima (Scolopacidae)

Few waders are as tightly restricted to one habitat as this one. It spends most of its time at the very edge of the surf, searching through weed-covered rocks for its food. It is hard to see against the dark seaweed, but its yellow-based, somewhat downcurved bill and yellow legs are useful clues to its identity.

FEEDS *on rocky shores in winter, and around piers and groynes. Breeds on northern tundra and mountains.*

very dark wings

scaly dark back

downcurved bill

yellow legs

VOICE *Simple, low, liquid* weet *or* weet-wit.
FEEDING *Takes a variety of insects, spiders, and other invertebrates in summer; chiefly periwinkles and similar molluscs in winter.*
DISTRIBUTION *Winters: Atlantic, south to northern Spain.*
SIMILAR SPECIES *Dunlin (p.193); the dull, winter Turnstone (above), with which it often flocks.*

Ringed Plover

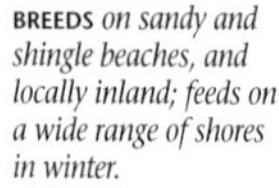

Charadrius hiaticula (Charadriidae)

BREEDS *on sandy and shingle beaches, and locally inland; feeds on a wide range of shores in winter.*

A small, pale plover with a striking head and breast pattern and bright orange legs, this bird typically feeds on sandy beaches in summer, or in tight flocks with other waders at high tide. Its eggs are well camouflaged against the background of sand or shingle.

VOICE *Characteristic fluty whistle, a bright, mellow* too-lit, *also a sharp* queep.
FEEDING *Picks small insects and worms from ground, using run-tilt action.*
DISTRIBUTION *Breeds: Atlantic, Baltic. Winters: Atlantic, locally Mediterranean.*
SIMILAR SPECIES *Kentish Plover (below); Little Ringed Plover* (C. dubius), *which has no wingbar.*

Kentish Plover

Charadrius alexandrinus (Charadriidae)

OCCUPIES *sandy areas near the coast around saline lagoons and salt pans, and on open sandy shores, especially during winter.*

Distinguished from adult Ringed Plovers (above) by its incomplete breastband and dark legs, the Kentish Plover is most common in the south of its range. Immature Ringed Plovers are similar, showing an incomplete breastband, but they are larger and have paler legs.

VOICE *Short, sharp, whistled* whip, *whistled* bew-ip; *rolled trilling notes.*
FEEDING *Small invertebrates from ground, caught in typical run-tilt plover action.*
DISTRIBUTION *Mediterranean and Atlantic, south from Denmark, although leaving more northerly parts of its range in winter.*
SIMILAR SPECIES *Juvenile Ringed Plover.*

Grey Plover

Pluvialis squatarola (Charadriidae)

A pale, silvery species in winter, with distinctive black "wing-pits", the Grey Plover is bigger than other plovers. It mostly feeds on mud- and sandflats, and roosts in flocks on fields and salt marshes. In summer its plumage changes, its matt black face, breast, and belly contrasting strongly with the piebald upperparts.

FEEDS *on large muddy estuaries and other shores from autumn to spring.*

pale eye-stripe

mottled grey

black wingpits

bulky, squat appearance

pale below

patchy plumage

LATE MOULTING

black underside

VOICE *High, plaintive* twee-oo-wee.
FEEDING *Worms, molluscs, and crustaceans picked and pulled from the upper layer of mud.*
DISTRIBUTION *Breeds: northern tundra. Winters: Atlantic and Mediterranean.*
SIMILAR SPECIES *Knot (p.193), which is similarly pale in winter, but is smaller; Golden Plover* (P. apricaria), *which has gold-spangled upperparts and white wingpits.*

Redshank

Tringa totanus (Scolopacidae)

Very conspicuous, due to its loud voice and bold white upperwing bands, the Redshank is common on many coasts but scarcer inland in areas where drainage has destroyed wet grassland. A wary bird, it flies off with noisy calls, alerting other birds to danger.

BREEDS *on salt marshes, wet pastures, and moors. Feeds on estuaries, and salt and freshwater marshes.*

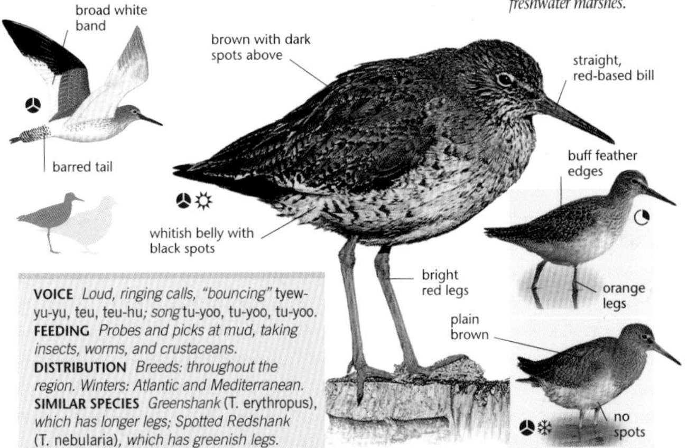

VOICE *Loud, ringing calls, "bouncing"* tyew-yu-yu, teu, teu-hu; *song* tu-yoo, tu-yoo, tu-yoo.
FEEDING *Probes and picks at mud, taking insects, worms, and crustaceans.*
DISTRIBUTION *Breeds: throughout the region. Winters: Atlantic and Mediterranean.*
SIMILAR SPECIES *Greenshank* (T. erythropus), *which has longer legs; Spotted Redshank* (T. nebularia), *which has greenish legs.*

Sanderling

Calidris alba (Scolopacidae)

FEEDS *in flocks on broad sandy beaches and estuaries; scarce on other shores and inland. Breeds on northern tundra.*

In winter, the Sanderling is by far the whitest of small waders. Typically, it occurs in small flocks at the water's edge on sandy shores, often in mixed groups with Ringed Plover (p.190) and Dunlin (p.193). In less typical standing water habitats, its active feeding style – snatching its food from the edge of waves – invites comparison with the smaller stints. However, unlike the Sanderling, these stints have more slender bills, which are shorter than the length of the head. In spring and autumn, the Sanderling's breast and back are mottled with chestnut, but its belly remains pure white, with a sharp demarcation line between the contrasting colours.

broad white wingbar

blackish patch on shoulder

plain, pale grey back

stout black bill

bright white underparts

relatively long legs

black spangled grey

NOTE

Sanderlings have a very unique style of feeding. Small groups dart back and forth along the edge of the waves, like mechanical toys, snatching items of food that is carried in by the surf. When they stop, which is very briefly, one can see their diagonistic lack of a hind toe.

VOICE *Sharp, hard, short* plit *or* twik twik.
FEEDING *Snatches small molluscs, marine worms such as sandhoppers and insects, as well as other invertebrates from the edge of waves.*
DISTRIBUTION *Winters: Mediterranean and on Atlantic shores, south from Denmark; all areas on migration.*
SIMILAR SPECIES *Dunlin (p.193); Curlew Sandpiper* (C. ferruginea), *which is larger and more elegant, with a longer bill and a white rump, and in summer has very different copper-red plumage; Little Stint* (C. minuta), *which is smaller, with a shorter bill.*

Dunlin

Calidris alpina (Scolopacidae)

Widespread and common on many European coasts, the Dunlin often occurs in large flocks, feeding and roosting together on mudflats and marshes. Its winter plumage is drab, but the thin white wingbar and white-sided, dark rump are distinctive. Summer birds are easily distinguished by the black belly patch and richly coloured back.

WINTERS *in estuaries and on open coasts; breeds in tundra, moorland, and northern coastal grassland.*

VOICE *Thin, reedy* shrree *or rasping* treerrr; *soft twitter from feeding flocks.*
FEEDING *Probes and picks worms, snails, and other invertebrates.*
DISTRIBUTION *Breeds: Atlantic, Baltic. Winters: Mediterranean, Atlantic.*
SIMILAR SPECIES *Knot (below); Sanderling (p.192); and Curlew Sandpiper* (C. ferruginea), *which is larger, with a white rump.*

Knot

Calidris canutus (Scolopacidae)

Marbled black, buff, and chestnut with coppery underparts, the Knot is among the most colourful of waders in spring and summer. In winter, it is a dull pale grey, but is still spectacular as it forms vast flocks of thousands that often take to the air, swooping through the sky in dramatic aerial manoeuvres.

ROOSTS *in dense flocks on muddy estuaries, and feeds on a wide variety of shores; breeds on tundra.*

VOICE *Rather quiet; dull, short* nut, *plus occasional bright, whistled note.*
FEEDING *Insects and plants in summer; molluscs, crustaceans, and worms in winter.*
DISTRIBUTION *Winters: Atlantic, locally in the Mediterranean; all areas on migration.*
SIMILAR SPECIES *Dunlin (above), which has a longer bill; Grey Plover (p.191), which has a shorter bill.*

Curlew

Numenius arquata (Scolopacidae)

WINTERS *mainly on big, muddy estuaries; breeds on bogs, wet moorland or meadows, and northern shores and islands.*

Europe's largest wader, the Curlew is widespread on all coasts and, especially when breeding, inland too. With its very long, downcurved bill, distinctive calls, and lovely song, it is hard to mistake for any common bird apart from a Whimbrel (*N. phaeopus*), although distant flying birds resemble gulls.

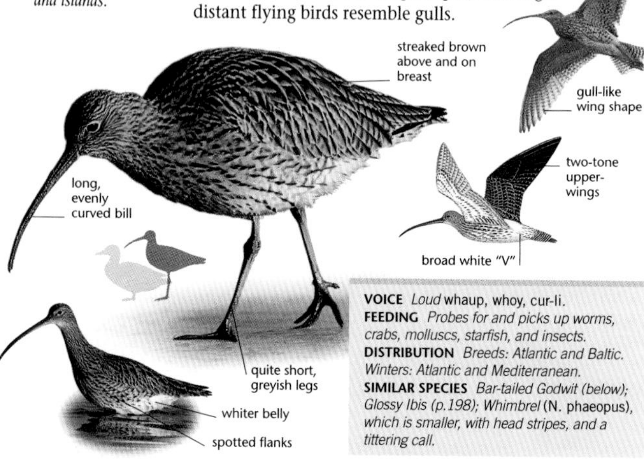

VOICE *Loud* whaup, whoy, cur-li.
FEEDING *Probes for and picks up worms, crabs, molluscs, starfish, and insects.*
DISTRIBUTION *Breeds: Atlantic and Baltic. Winters: Atlantic and Mediterranean.*
SIMILAR SPECIES *Bar-tailed Godwit (below); Glossy Ibis (p.198); Whimbrel* (N. phaeopus), *which is smaller, with head stripes, and a tittering call.*

Bar-tailed Godwit

Limosa lapponica (Scolopacidae)

WINTERS *on broad estuaries and sheltered muddy and sandy beaches; rarely inland.*

Although it breeds only on the Arctic tundra, the Bar-tailed Godwit is far more widespread on the coasts of Europe in winter than the larger Black-tailed Godwit (p.195). Flocks disperse to probe for food in the mud; these often have a habit of rolling and twisting as they fly in to roost at high tide.

plain upperwings with dark tips

barred tail

streaked grey-brown and buff

long, slightly upcurved bill

1ST

quite short, dark legs

♂

coppery red below

streaked bright buff

VOICE *Rapid, yelping* kirruk kirruk *flight call.*
FEEDING *Probes in mud and sand for large marine worms and molluscs.*
DISTRIBUTION *Winters: Atlantic, south from Scotland; locally in the Mediterranean.*
SIMILAR SPECIES *Curlew (above), which has a downcurved bill; Black-tailed Godwit (p.195), which is more elegant, with longer neck and legs, and a long, straight bill.*

Black-tailed Godwit

Limosa limosa (Scolopacidae)

A large, handsome wader, the Black-tailed Godwit is unmistakable in flight, when it reveals its boldly pied wing and tail pattern. The only possibility of confusion is with an Oystercatcher (p.188), but the average bill and black head of the latter are obvious, even at a distance. At rest, however, it could be confused with the Bar-tailed Godwit (p.194), although it has a longer, straighter bill, longer legs, and a more elongated appearance. The two species also have different habitat preferences within the winter feeding ranges – the Black-tailed Godwit favours deep, sloppy mud, whereas the Bar-tailed Godwit prefers sandier surfaces.

BREEDS *in wet meadows and flooded pastures; winter flocks prefer narrow, sheltered estuaries with long strips of rich mud.*

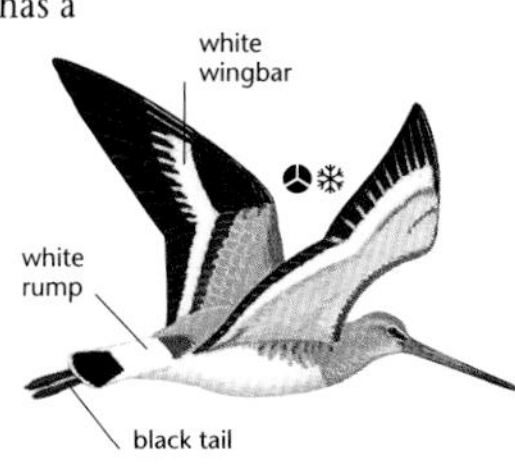

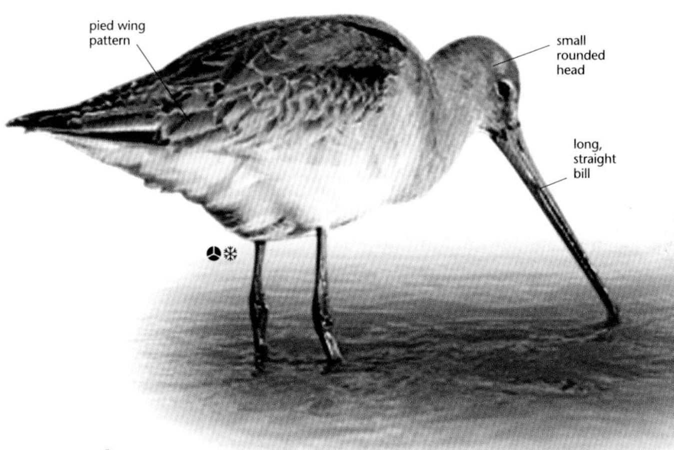

NOTE

The Icelandic, Scottish, and north Norwegian breeding birds (subspecies L. l. islandica*), which winter in western Europe, have shorter legs and bills, but are more richly coloured in summer, than the southerly breeders. These winter from the Mediterranean southwards.*

VOICE *Frequent nasal* weeka-weeka-weeka *calls when breeding; rapid* vi-vi-vi *flight calls.*

FEEDING *Probes deeply for worms, molluscs, and seeds, often wading up to belly in water.*

DISTRIBUTION *Breeds: Atlantic, north from France; Baltic. Winters: Atlantic, south from Scotland; locally in the Mediterranean.*

SIMILAR SPECIES *Oystercatcher (p.188), which has a similar flight pattern, but its black and white plumage and orange bill are very distinctive; Bar-tailed Godwit (p.194), which is altogether less elegant.*

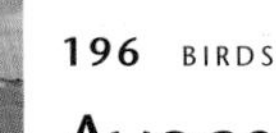

Avocet

Recurvirostra avosetta (Recurvirostridae)

WINTERS *in close flocks that fly and feed on muddy estuaries. Breeds on shallow, saline coastal lagoons and by muddy pools.*

The Avocet is a distinctive wader, handsome and graceful, with a strongly upturned bill, which it sweeps from side to side through shallow water in search of food. Conservation and habitat management have helped provide it with its very specific needs: shallow, brackish water and oozy mud for feeding, and drier islands for nesting. As a result, it has thrived and spread. Avocets can be surprisingly difficult to spot among flocks of the similarly sized Black-headed Gull (p.173), especially when roosting with their head and bill under a wing.

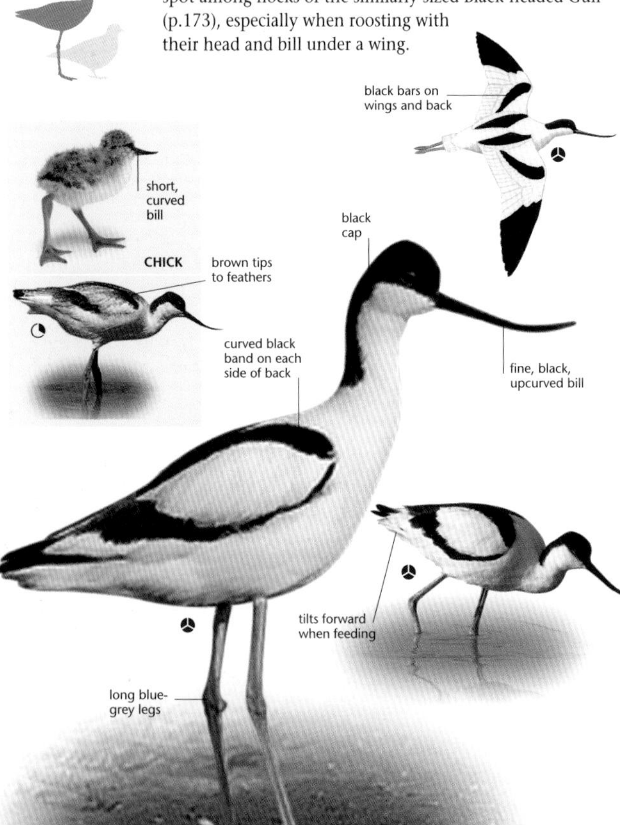

VOICE *Loud, fluty* klute *or* kloop.
FEEDING *Sweeps upcurved bill sideways through shallow water to detect and snap up tiny shrimps, worms, and other invertebrates.*
DISTRIBUTION *Breeds: Mediterranean; Atlantic, north to southern Norway; south Baltic. Winters: Atlantic, south from Britain; and Mediterranean.*
SIMILAR SPECIES *Black-winged Stilt (p.197); Black-headed Gull (p.173), especially if backlit (sun is behind the bird from viewer's position) and roosting with their heads under their wings.*

NOTE

Not a particularly hardy bird, the Avocet has recently spread northwards in its winter distribution, many birds now remaining around the North Sea. This range extension has been attributed to climate change.

Black-winged Stilt

Himantopus himantopus (Recurvirostridae)

This elegant, jet-black and dazzling white wader has the longest legs relative to body length of any of the world's birds. This enables it to wade into deep water in search of food. Essentially unique in Europe, its centre of distribution is the Mediterranean region, where it breeds by shallow fresh, brackish, or salt waters.

NESTS *on salt pans, coastal lagoons, reedy ponds, and flooded fields; also on open shores on migration.*

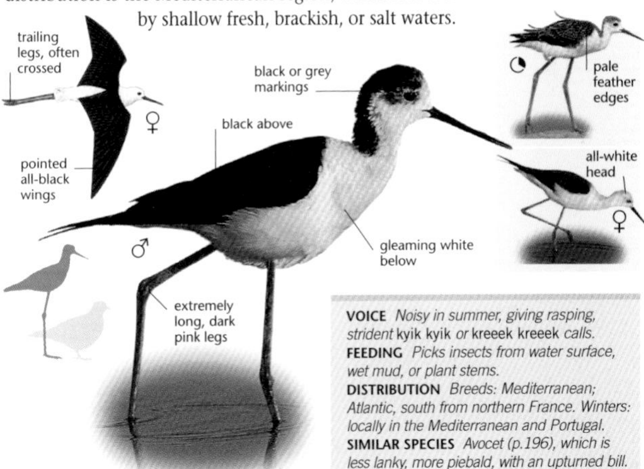

VOICE *Noisy in summer, giving rasping, strident* kyik kyik *or* kreeek kreeek *calls.*
FEEDING *Picks insects from water surface, wet mud, or plant stems.*
DISTRIBUTION *Breeds: Mediterranean; Atlantic, south from northern France. Winters: locally in the Mediterranean and Portugal.*
SIMILAR SPECIES *Avocet (p.196), which is less lanky, more piebald, with an upturned bill.*

Collared Pratincole

Glareola pratincola (Glareolidae)

An unusual, specialized wader, the Collared Pratincole has a swallow-like form related to its aerial feeding habits. It is basically a Mediterranean bird that sometimes strays farther north. Despite its elegance in flight, it can look dumpy on the ground with its feathers fluffed up. It often feeds in small groups.

INHABITS *mainly extensive areas of flat, dry mud, damp pasture, drained marshes, salt pans, and bare ground.*

VOICE *Sharp, far-carrying, tern-like* ki, kitik, *rhythmic* kirri-tik-kit-ik.
FEEDING *Catches airborne insects in its bill while flying.*
DISTRIBUTION *Mediterranean, Atlantic (southern Iberia).*
SIMILAR SPECIES *Black-winged Pratincole* (G. nordmanni), *which has black underwings, and is a migrant in the eastern Mediterranean.*

Glossy Ibis

Plegadis falcinellus (Threskiornithidae)

A uniformly dark, long-legged water bird, with a downcurved bill, the Glossy Ibis is unmistakable, apart from the potential confusion with a Curlew (p.194), especially when it is backlit. In flight, its neck and legs are extended, which coupled with the flight action – mechanical wingbeats, interspersed with short glides – gives the bird a distinctive appearance.

NESTS *in trees and reedbeds, around shallow coastal lagoons and deltas.*

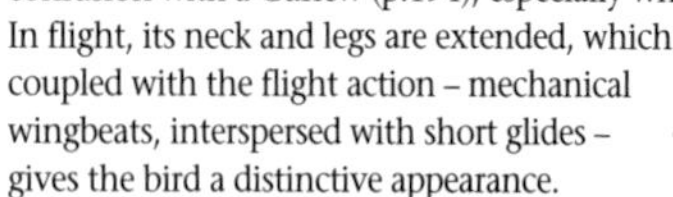

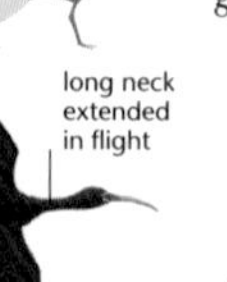

VOICE *Harsh, guttural squawks near the nest.*
FEEDING *Insects, molluscs, other invertebrates, small amphibians, and fish.*
DISTRIBUTION *Breeds: eastern Mediterranean. Winters: Atlantic (Spain and Portugal) and locally Mediterranean.*
SIMILAR SPECIES *Curlew (p.194); flight action resembles that of the Pygmy Cormorant* (Phalacrocorax pygmaeus).

Greater Flamingo

Phoenicopterus ruber (Phoenicopteridae)

The only one of the world's five species of flamingo to occur in the wild in Europe, the Greater Flamingo is one of our most exotic-looking birds. This gregarious bird has an angled-down bill, extremely long neck and legs, and pastel plumage.

BREEDS *in some Mediterranean salty lakes; a non-breeder on salt pans, lagoons, and deltas.*

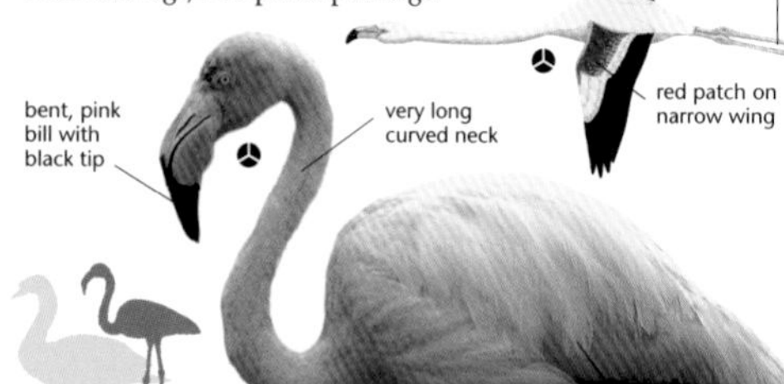

VOICE *Loud, deep honking and cackling.*
FEEDING *Sweeps bill upside down through water to catch tiny crustaceans.*
DISTRIBUTION *Breeds: Mediterranean (very localized). Winters: Mediterranean and Atlantic (Portugal).*
SIMILAR SPECIES *White Stork* (Ciconia ciconia), *which is similarly large, but is white with black flight feathers, and a straight red bill.*

Little Egret

Egretta garzetta (Ardeidae)

A dazzling white water bird, the Little Egret is the most widespread of Europe's white herons. Found around both coastal and inland wetlands, it has been steadily extending its range northwards, up the Atlantic coast, for the past fifty years. Sometimes wading slowly or standing still, it is often active and agile in search of prey. In flight, its yellow feet should provide a disinct identification feature, except that they are often caked with mud. Little Egrets are communal birds, both nesting and roosting in colonies in waterside trees, especially around coastal lagoons and marshes.

INHABITS *a range of coastal wetlands, from lagoons and salt marshes to rocky shores; also around wetlands inland.*

VOICE *Usually silent except when breeding, when it utters snarling and croaking calls.*
FEEDING *Eats small fish, frogs, and snails.*
DISTRIBUTION *Mediterranean; Atlantic, north to Britain.*
SIMILAR SPECIES *Other egrets, such as the Great White and Cattle Egret (see note); Spoonbill* (Platalea leucorodia), *which has a spoon-shaped bill; Squacco Heron* (Ardeola ralloides), *which appears dark buff at rest, but reveals surprisingly pure white wings and tail in flight, resembling an egret.*

NOTE

As the commonest species among egrets, the Little Egret is the benchmark against which we can compare other large white water birds: Great White Egret (E. alba) *has a larger yellow bill; Cattle Egret* (Bubulcus ibis) *looks dumpy, and is buff on the head.*

House Martin

Delichon urbica (Hirundinidae)

BUILDS *a mud nest on rock cliffs, under an overhang; buildings are now the predominant breeding habitat.*

Small and stocky, with pied plumage and a bold white rump, the House Martin is familiar in towns and villages. Its natural breeding sites, however, are hard cliffs, both coastal and inland. It feeds entirely in the air on small flies and similar prey, circling over rooftops or low over fresh waters. It comes to the ground to gather mud, which it uses to build its distinctive nest.

VOICE *Hard, quick, chirping* prrit *or* chrrit, tchirrup; *twittering song of similar notes.*
FEEDING *Catches insects in flight, high up.*
DISTRIBUTION *Throughout, in summer; a few by Mediterranean coasts in winter.*
SIMILAR SPECIES *Sand Martin (below); Storm Petrel* (Hydrobates pelagicus), *which is almost all black, with a white rump. It is a similar size, but almost always seen at sea.*

Sand Martin

Riparia riparia (Hirundinidae)

EXCAVATES *nest in holes in sandy cliffs, both coastal and inland, natural and man-made; usually near water.*

The smallest of the European swallows and martins, with the most fluttering flight, the Sand Martin is the first to appear on its northern breeding grounds in spring. At this time, it usually hunts over water, where it can rely on a supply of flying insect prey. Always gregarious, it roosts in reedbeds.

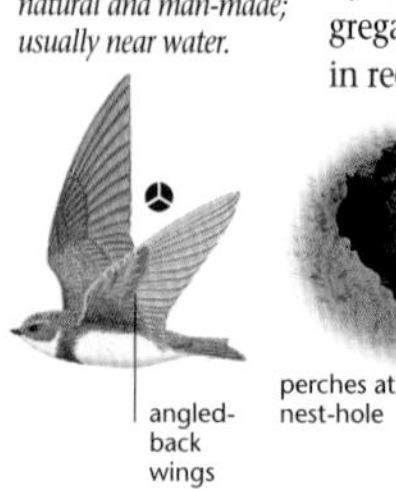

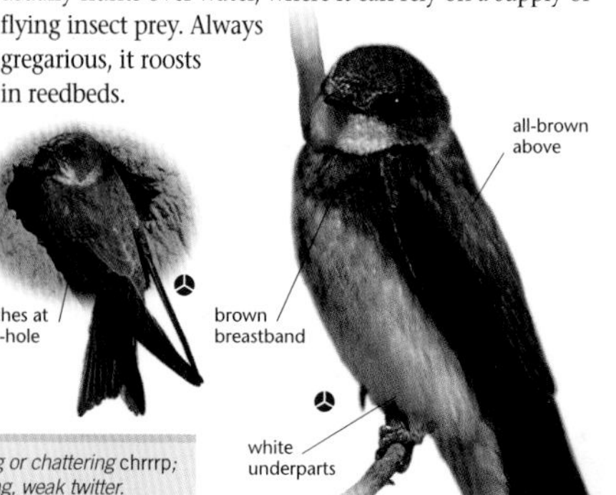

VOICE *Low, dry, rasping or chattering* chrrrp; *song rambling, chattering, weak twitter.*
FEEDING *Catches insects in flight, often over water; sometimes feeds on bare ground.*
DISTRIBUTION *Throughout in summer.*
SIMILAR SPECIES *House Martin (above); Swallow* (Hirundo rustica), *which is blue-black above, with long tail streamers; Crag Martin* (Ptyonoprogne rupestris), *which is more dusky.*

FEEDS *in winter on salt marshes and grazing marshes.*

Twite

Carduelis flavirostris (Fringillidae)

A drab-looking finch, especially during the winter, when the Twite is most likely to be seen on the coast. It is the upland and northern counterpart of the familiar, more richly coloured Linnet (*C. cannabina*). Largely brown and grey, the male has a pinkish rump and greyish streaks on the wings, both visible in flight. It is the call, after which it is named, that clinches its identification.

VOICE *Nasal, rising* twa-eet, *chattering flight call, and song a twittering warble.*
FEEDING *Eats small seeds.*
DISTRIBUTION *Breeds: Atlantic, north from Ireland in mountains and northern coasts. Winters: Atlantic, southern Baltic.*
SIMILAR SPECIES *Linnet* (C. cannabina), *which has brighter plumage, with more red; they often occur together in mixed flocks.*

Shorelark

Eremophila alpestris (Alaudidae)

A long, sleek lark with a unique head pattern that is boldest in summer, the Shore Lark breeds in the mountains of Scandinavia and southeast Europe. In winter, it appears on beaches and marshy spots around the Baltic and North Seas, where it often feeds alongside Snow Buntings (p.202).

FEEDS *on sandy beaches near the high tide mark in winter, and nearby marshes and fields. Breeds on mountains.*

VOICE *Call pipit-like, thin* tseeeep *or louder* seep-seep.
FEEDING *Creeps over ground, taking seeds, insects, crustaceans, and tiny molluscs.*
DISTRIBUTION *Winters: Atlantic, from northern France to Denmark; southern Baltic.*
SIMILAR SPECIES *Rock Pipit (p.203); Skylark* (A. arvensis); *other larks around the Mediterranean.*

Snow Bunting

Plectrophenax nivalis (Emberizidae)

FEEDS *on shingle banks and muddy coastal marshes in winter, also exposed mountain slopes.*

The stark black and white of the breeding male Snow Bunting is well suited to its snowy northern breeding habitat. However, over much of Europe it is seen only in its subdued winter plumage, when the sexes look more alike. They have a distinctive low-slung look on the ground due to their short legs. Feeding flocks move forward by leap-frogging flights, the birds at the back flying over those in front.

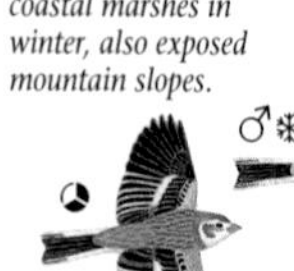

VOICE *Call loud, deep, clear* pyiew *or* tsioo, *frequently lighter trilling.*
FEEDING *Seeds and invertebrates picked from strandlines in winter; insects in summer.*
DISTRIBUTION *Winters: Atlantic, France to Scotland; southern Baltic. On passage, Norway and Baltic.*
SIMILAR SPECIES *Lapland Bunting (below), which shows no white trace on wings in flight.*

Lapland Bunting

Calcarius lapponicus (Emberizidae)

FEEDS *mainly on salt marshes and short, wet grassland near coasts in winter. Breeds on northern tundra.*

Rarely seen in summer, when it breeds in remote, wild places, the Lapland Bunting appears on more southerly coasts in autumn and winter. It keeps a low profile, creeping among the grasses on dunes, salt marshes, and golf courses, and usually stays unnoticed until flushed from underfoot.

mainly rufous head

dark ear coverts

dark tail with white sides

black cap, face, and breast

VOICE *Call a staccato rattle and clear whistle.*
FEEDING *Seeds and invertebrates picked from the ground, especially in summer.*
DISTRIBUTION *Winters: Atlantic, from Brittany to Scotland and Denmark. On passage, Norway and Baltic.*
SIMILAR SPECIES *Snow Bunting (above); Corn Bunting* (Miliaria calandra), *which is bigger and plainer.*

Rock Pipit

Anthus petrosus (Motacillidae)

For much of the year, the habitat of the Rock Pipit helps to betray its identity, for it is truly a bird of rocky coasts and islands. Its song-flight and song are similar to those of other pipits, but its loud, single call note is easily recognizable. The Scandinavian and Baltic birds, race *A. p. littoralis*, are especially distinctive, with a strong eye-stripe, blue-grey head, and pinkish chest in summer.

BREEDS *on rocky coasts; on migration and in winter, some visit salt marshes and soft shores, fewer inland by water.*

blue grey tone

more defined eye shape

dark tail

SCANDINAVIAN RACE

dark back

dull below

dark legs

sometimes grey nape

weak stripe over eye

diffuse streaks on dusky olive back

blurry dark streaks

grey tail feathers

dull white to yellowish underside

dark legs

long claws

VOICE *Call a slurred* pseeep; *song a descending cadence, in song-flight.*
FEEDING *Forages mainly on cliff-tops in summer, on stony/seaweedy shores in winter, for insects, sandhoppers, and small molluscs.*
DISTRIBUTION *Breeds: Atlantic, north from France; Baltic. Winters: Atlantic, and locally Mediterranean.*
SIMILAR SPECIES *Meadow Pipit* (A. pratensis), *which is more olive above; Water Pipit* (A. spinoletta), *which is more colourful than the Scandinavian and Baltic subspecies* A. p. littoralis, *and breeds only in mountains, but often winters on coastal lagoons and marshes.*

NOTE

Until recently, Rock Pipits and Water Pipits (A. spinoletta) *were considered as coastal and montane forms of one species. They mix in coastal areas in winter, where they can be distinguished by the colour of their outer tail feathers, which are grey and white, respectively.*

Short-eared Owl

Asio flammeus (Strigidae)

HUNTS *over coastal grazing marshes, especially in winter; breeds on heaths, moors, young plantations, and sporadically in rough coastal areas.*

One of the few owls that regularly appears in broad daylight, the Short-eared Owl is often seen hunting like a harrier. With its wings held in a very shallow "V", it flies and glides low over open grassland, heaths, and coastal marshes. Its dark-rimmed, yellow eyes give it a fierce expression that is often visible at long range. It has a strongly barred tail and white trailing edge to the wings. Its numbers fluctuate markedly in response to cyclical variation in the numbers of voles, which is its favoured food.

NOTE

Other large, day-flying predators, which may be seen around grazing marshes and other coastal habitats in winter include Hen and Marsh Harriers (Circus cyaneus) *and* (C. aeruginosus), *and Rough-legged Buzzard* (Buteo lagopus). *The longer-tailed harriers quarter the marshes with wings held in a shallow "V", while the buzzard habitually hovers in search of prey.*

VOICE *Nasal bark; male's song a soft booming hoot, given in display flight, accompanied by wing-claps.*
FEEDING *Small mammals, especially voles; some birds caught in flight.*
DISTRIBUTION *Breeds: Atlantic, north from France, and Baltic. Winters: Mediterranean, and Atlantic, south from Britain and Denmark.*
SIMILAR SPECIES *Long-eared Owl* (A. otus), *which is sometimes seen by day, and has a less gliding flight and a more faintly barred tail; female Hen Harrier* (Circus cyaneus), *which shows a white rump in flight.*

Merlin

Falco columbarius (Falconidae)

A small, dynamic falcon of open country, the Merlin flies fast and low over the ground in pursuit of prey, with rapid flicks of its wings and a final agile rise to strike. The thrush-sized male is bluish grey above with a dark tail-band, while the bigger female is earthy brown with a cream-barred tail.

FREQUENTS *coastal grazing marshes and extensive salt marshes in winter; breeds in northern moorland, tundra, and young conifer plantations.*

VOICE *Quiet when away from the nest and outside the breeding season.*
FEEDING *Small birds, caught in flight, such as larks, pipits, and small waders; large insects.*
DISTRIBUTION *Winters: north Mediterranean, Atlantic, southern Baltic.*
SIMILAR SPECIES *Peregrine (below); Hobby* (F. subbuteo), *which has strong face markings, red thighs, and a swift-like outline.*

Peregrine

Falco peregrinis (Falconidae)

The big, powerfully built Peregrine is a bird-killing falcon. It uses its famous high-speed diving "stoop" to kill its prey in mid-air. When hunting, it often patrols at great heights, looking like a tiny black anchor in the sky. Its arrival causes confusion and panic among bird flocks in estuaries.

HUNTS *over estuaries and marshes in winter; breeds on rocky cliffs, and on large buildings.*

VOICE *Loud, raucous calls at nest, throaty* haak-haak-haak-haak.
FEEDING *Birds, the size of a small wader to a duck, killed after chase or dramatic dive.*
DISTRIBUTION *Throughout, except northern regions in winter; Baltic, only on passage.*
SIMILAR SPECIES *Merlin (above), Gyrfalcon (p.206), and Kestrel* (F. tinnunculus), *which is more slender with a distinctive hovering action.*

Gyrfalcon

Falco rusticolus (Falconidae)

BREEDS *on rocky cliffs, often near seabird colonies, also inland; hunts over tundra and lower coastal areas in winter.*

The largest of all the falcons, Gyrfalcon is a northern species, often breeding on cliff ledges, in the old nests of other birds. Its huge build and large wingspan, 1.1–1.3m, the same as a Buzzard (*Buteo buteo*), are unmistakable. It occurs in a range of colour forms, from dark (like a dull Peregrine) to almost pure white, paler forms, found at higher latitudes.

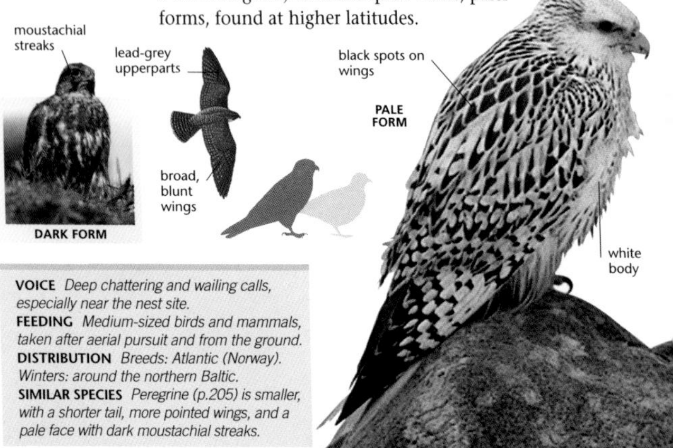

VOICE *Deep chattering and wailing calls, especially near the nest site.*
FEEDING *Medium-sized birds and mammals, taken after aerial pursuit and from the ground.*
DISTRIBUTION *Breeds: Atlantic (Norway). Winters: around the northern Baltic.*
SIMILAR SPECIES *Peregrine (p.205) is smaller, with a shorter tail, more pointed wings, and a pale face with dark moustachial streaks.*

Eleonora's Falcon

Falco eleonorae (Falconidae)

NESTS *on cliffs and rock ledges, usually close to the sea; feeds actively over the sea and coastal areas.*

A summer visitor to the Mediterranean region, Eleonora's Falcon feeds on small birds, and its breeding is timed to coincide with autumn migration. Of all the European falcons, it has the most agile flight. It occurs in two colour forms: the dark form is slaty grey; the pale form has buff underparts, streaked with black, and a distinct white face.

VOICE *Repetitive harsh cries, especially around breeding sites.*
FEEDING *Mainly small birds, especially migrants; also large insects, caught in flight.*
DISTRIBUTION *Mediterranean.*
SIMILAR SPECIES *Peregrine (p.205) is larger; Hobby* (F. subbuteo) *is more richly coloured than the pale form; Red-footed Falcon* (F. vespertinus) *has red thighs.*

Sea Eagle

Haliaeetus albicilla (Accipitridae)

HUNTS *on rocky coasts, estuaries, and remote marshes in summer; winters mainly on large, damp coastal plains.*

Also known as White-tailed Eagle, the Sea Eagle is a huge, heavy-billed bird, with very long, broad, fingered wings and a short tail, producing a distinctive flight silhouette. Adults have pale heads, yellow bills, and striking white tails, but young birds are darker overall, lacking the white wing patches of the similarly-sized, but longer tailed, immature Golden Eagle (*Aquila chrysaetos*). A much persecuted bird, the Sea Eagle is now being restored to parts of its former range through reintroduction schemes, using stocks from Norway where it is locally abundant.

VOICE *Shrill yaps near nest in summer.*
FEEDING *Fish and fish offal; seabirds and mammals, up to the size of a hare; carrion.*
DISTRIBUTION *Breeds: eastern Mediterranean; Atlantic, north from Scotland; Baltic; more widespread on western European coasts in winter.*
SIMILAR SPECIES *Griffon Vulture* (Gyps fulvus), *which is another "flying barn-door", and is found locally in the Mediterranean cliff colonies. Golden Eagle* (Aquila chrysaetos), *which has narrower wings and a longer tail, and is usually more associated with upland habitats.*

NOTE

Another large, fish-eating raptor, which feeds in estuaries and sometimes breeds on cliffs, is the Osprey (Pandion haliaetus). *This bird is dark above and pale below, and is altogether a less bulky bird than the Sea Eagle. It is often seen hovering before it dives.*

Strandline finds

The strandline opens a window on worlds that are otherwise inaccessible to land-based observers. Sea currents may carry plants and animals that have drifted from the oceans depths or distant shores, and wash ashore life from nearby waters – sublittoral life that is otherwise outside the scope of this book. The strandline also gives an insight into the past, with rocks and fossils released from land by coastal erosion, and helps build knowledge of the history of the Earth. While strandline finds are fascinating, some care is needed. Sea-borne debris is inevitably accompanied by human litter from across the world. A beachcomber should wear stout footwear and gloves for protection against sharp glass and plastic, toxic chemicals, stinging jellyfish, and other dangerous finds.

SEA BALLS

BUOY-BARNACLE

PELICAN'S FOOT

LEATHERBACK TURTLE

Plants

The remains of vegetation are usually the most significant component of a strandline, and may comprise seaweeds and other marine plants, as well as the trunks, branches, and seeds of terrestrial plants that have found their way into the sea. This stranded plant material is an important food resource for numerous invertebrate decomposers, such as beach-fleas and sand-hoppers (p.139), which in turn provide rich pickings for birds such as the Turnstone (p.189).

FOUND *on all intertidal habitats, at the furthest reach of the high tide; strandlines are best developed on sheltered shores.*

▶ **SEA BALLS** *are also known as Nuns' Farts in the Balearics, and are a familiar sight by the Mediterranean. Up to 8cm wide, they are the fibrous remains of the leaves and rhizomes of Neptune Weed (p.54), tumbled into a ball by the action of waves.*

▼ **BEACH LITTER** *consists of fragments of plants, seaweeds, or sea-grasses, along the high tide mark, which provide food for invertebrates and, as they rot, nutrients for terrestrial plants.*

remains of intertidal and sublittoral seaweeds

▼ **SEA BEANS** *are seeds of tropical American trees and vines, carried to Atlantic shores by the Gulf Stream and North Atlantic Drift. They remain viable even after days of drifting.*

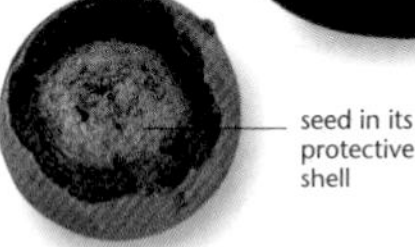

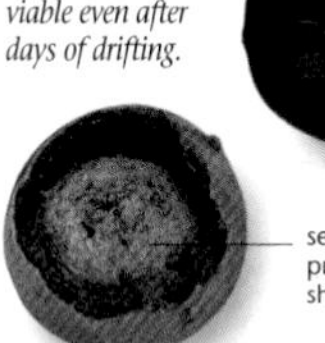

NOTE

It is said Columbus found Sea Beans in Galway and inferred there was land across the ocean, leading to discovery of the Americas.

▼ **KELP HOLDFASTS** *anchor seaweeds to rocks in the face of battering waves. A beached example demonstrates the complex structure, which in situ forms a protected microhabitat for numerous animals and plants.*

Invertebrates

FOUND *in all marine drift habitats, but especially close to the habitat where they occur when alive.*

Invertebrates restricted to the sublittoral zone, below the extreme low-water mark, are mostly outside the scope of this guide. However, a strandline can, especially after high tide or after a storm, reveal a tantalizing glimpse of the underwater world that is otherwise known only to divers. Dead or alive, specimens that are deposited on the beach by the sea provide a rich food resource for a variety of species, making the strandline a valuable habitat for scavenging invertebrates and birds, especially gulls (pp.170–73).

fur-like scales

▲ **SEA-MOUSE** (Aphrodita aculeata) *is a worm up to 20cm long and is covered in fur-like scales above. Underneath, it is ribbed like the sole of a shoe. Its scales have a blue or bronze iridescence.*

8–14 arms

spiny fringe

▶ **COMMON SUN-STAR** (Crossaster papposus) *is distinguished from other starfish (pp.108–09) by its larger, and more variable, number of arms, and a coarse, spiny upper surface.*

5 points fan out from aperture margin

▼ **TUSK-SHELL** (Antalis vulgaris) *and related molluscs live buried, broad end downwards, in offshore fine sands and gravels. This tusk-shell has a cylindrical, tapering shell, to 6cm long.*

▲ **PELICAN'S FOOT** (Aporrhais pespelecani) *is a mollusc of up to 4cm long. It is named after its characteristic aperture margin, which has five points arranged in a fan shape.*

whorled spire

opaque white shell

porous structure

◀ **CUTTLEFISH BONES** *form the internal shell of cuttlefish (usually* Sepia officinalis*), which is a free-swimming, tentacled mollusc related to the squid and octopus. The bones are up to 20cm long.*

◀ **SEA WASH BALLS** *are the spongy masses, up to 10cm wide, of egg capsules of a Common Whelk (p.115). Each cell contains several eggs, although usually only one survives to emerge as the shellfish.*

▶ **SEA CHERVIL** *are rubbery masses of the bryozoan* Alcyonidium diaphanum *that are washed up on the shore. They occur in a variety of lobed and branched forms, the lobes and branches arising from a narrow base, which in life is attached to a stone.*

NOTE

Sublittoral invertebrates are usually washed up after a storm, which produces powerful breakers that can scour the sea bed.

broad,
flattened
fronds

▶ **HORNWRACK** (Flustra foliacea) *is another bryozoan, with a flattened branching structure. It is a familiar driftline stranding, washed ashore from living colonies.*

▼ **A SEA-URCHIN TEST** *is the empty, spineless shell of the sea-urchin, usually identifiable by its size, colour, profile, and surface sculpturing. This is a test of the Edible Sea-urchin (p.110).*

Ocean Drifters

LIVING *mainly in warm oceanic water, jellyfish are most often stranded in the summer months, sometimes in great numbers.*

The ocean may not seem to hold many opportunities for life, but the strandlines can demonstrate otherwise. Some of the ocean-drifting invertebrates featured here can be very abundant, especially after sustained periods of onshore winds, which carry surface-floating species for considerable distances. Sea currents do the same for creatures that inhabit the water column.

▲ **COMMON GOOSE-BARNACLE** (Lepas anatifera) *hitches a ride on all kinds of ocean debris by attaching itself to it with a stalk, to 90cm long. The "head" is covered with white plates on a dark brown skin.*

▼ **BY-THE-WIND-SAILOR** (Velella velella) *is a colonial hydrozoan, each colony having a chambered float and an erect sail. Fresh specimens are a transparent blue, but they soon decay to the chitinous float and sail.*

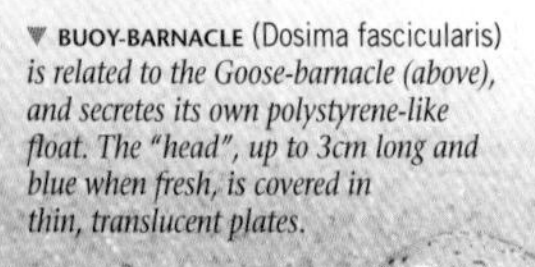

◀ **PORTUGUESE MAN-O'WAR** (Physalia physalis) *is a colonial hydrozoan with many tentacles, each several metres long and armed with dangerous stinging cells.*

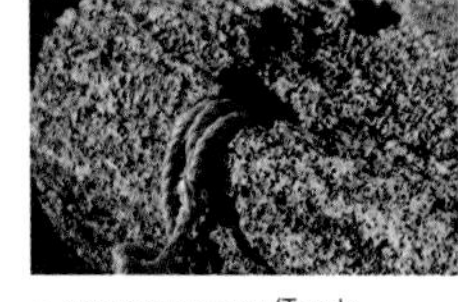

▲ **GREAT SHIPWORM** (Teredo navalis) *is a mollusc with a ridged shell. It bores into waterlogged driftwood, secreting a chalky substance that lines the borings.*

▼ **BUOY-BARNACLE** (Dosima fascicularis) *is related to the Goose-barnacle (above), and secretes its own polystyrene-like float. The "head", up to 3cm long and blue when fresh, is covered in thin, translucent plates.*

keeled plates

float

▲ **COMPASS JELLYFISH** (Chrysaora hysoscella) *is so called because of the radiating, branched colour pattern on its bell. It has 24 long tentacles, which hang from the rim; both the tentacles and the bell have stinging cells.*

▼ **BLUE JELLYFISH** (Cyanea lamarckii) *is usually violet or blue in colour. It is up to 30cm wide, with a lobed margin, and has many tentacles that are borne in groups. The tentacles of the Blue jellyfish do not sting strongly.*

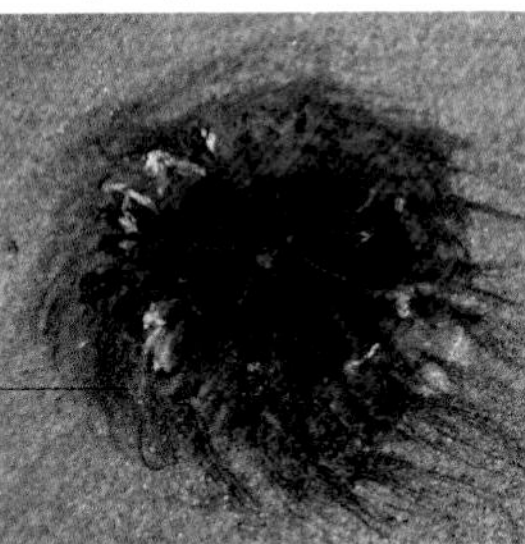

▶ **RHIZOSTOMA PULMO** *is a largely Mediterranean jellyfish, with a solid bell up to 50cm wide. It lacks tentacles and stinging cells. It has eight elongated lobes, with numerous mouths, which project from the underside, enabling it to eat planktonic food.*

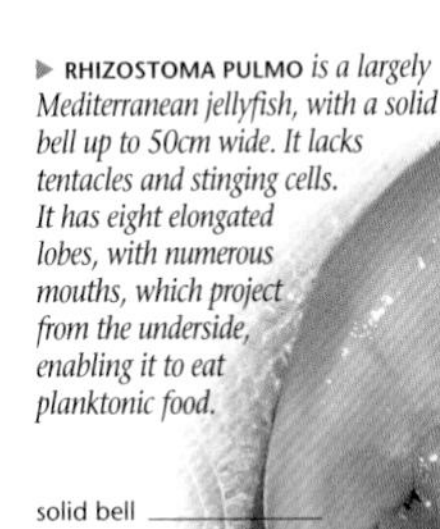

◀ **MOON JELLYFISH** (Aurelia aurita) *is a common swarming jellyfish, usually identifiable by its reproductive rings. It is frequently encountered in estuaries in late summer.*

▶ **SEA-GOOSEBERRY** (Pleurobrachia pileus), *is a member of the* Ctenophora *or comb-jelly family. It has a transparent, oval, jelly body. In water, its tentacles catch plankton on their sticky cells.*

NOTE

Stinging cells on the tentacles of jellyfish and Portuguese Man-o'war remain active even when the animal is stranded, so one should be careful of these.

Vertebrates

WASHED *onto a beach, any dead vertebrate will attract scavengers – look for a feeding frenzy among the gulls.*

Few fish are featured in this guide, as they are mostly not visible, except when beached. After a storm tide, strandlines can provide a cross-section of this hidden biodiversity, but one should aim to get out early, before they are consumed by scavenging gulls and crows. Other marine vertebrates, including whales and dolphins, are likely to last longer because of their size; some species are known to be from European waters only on the basis of strandings.

▼ **MERMAID'S PURSES** *are egg cases of fish related to sharks. This example, with a curving spike at each corner, comes from a skate or a ray.*

NOTE

Fish from warmer, southern waters are increasingly being found on European strandlines, tangible evidence that our climate is changing.

▶ **LESSER-SPOTTED DOGFISH** (Scyliorhinus caniculus) *is a small relative of the sharks. This dogfish is a common species in shallow water over soft sediments, and is up to 75cm long.*

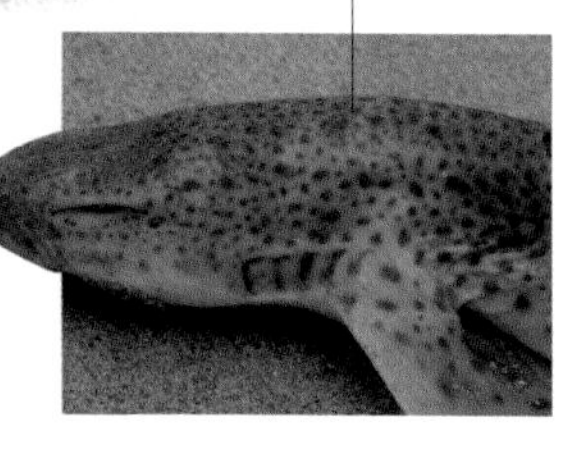

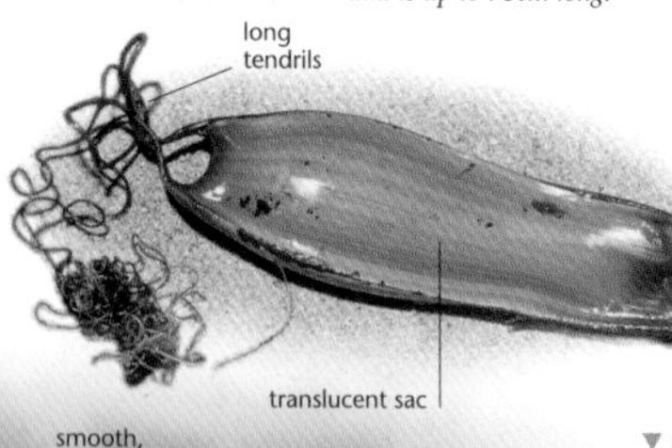

◀ **A MERMAID'S PURSE** *that is freshly laid is olive-brown and translucent. The long, curly tendrils, which help it to stay attached to seaweeds, indicate that this one was produced by a dogfish.*

▼ **LEATHERBACK TURTLE** (Dermochelys coriacea) *is a true oceanic wanderer and a vagrant to European shores. It is up to 2.9m long, and has a ridged, leathery shell that is very different from the shell of horny plates shown by all related species.*

Glossary

Words in *italics* are defined elsewhere in the glossary.

ACHENE A dry, one-seeded, non-splitting fruit, often with a *pappus*.

ACROCARPOUS Describes a moss, often erect in growth, the shoot sometimes forming dense cushions, in which the fruit capsules arise from the tips of the shoots or a major branch.

ACRORHAGI A wart-like structure, often brightly coloured, that forms a ring around the mouth of some sea-anemones.

ADULT In birds, a fully mature individual, able to breed, showing the final plumage pattern that no longer changes with age.

AESTIVATE To pass hot, dry summer conditions in a state of torpor, the summer equivalent of winter hibernation.

AGGREGATE (AGG.) A group of closely related species, usually difficult to tell apart, and often of uncertain taxonomic status.

ALTERNATE Borne singly, and alternating in two vertical rows or spirally.

AMPHIPODS A group of small, shrimp-like *crustaceans*, flattened from side to side, which live mainly on the sea floor and feed on *detritus*.

ANTHER The male part of a flower that bears the pollen.

ARTHROPODS A major group (*phylum*) of invertebrate animals with jointed legs and a hard outer skeleton. It includes *crustaceans*, insects, and spiders.

AWN A stiff, bristle-like projection from the individual flowers of some grasses.

AXIL The angle between two structures, such as a leaf and stem or the *midrib* and a small vein on a leaf.

BARNACLES Specialized *crustaceans*, the adults of which live attached to rocks and other surfaces. They are protected by hard shell-like plates, and *filter-feed* using highly modified limbs.

BARRED With marks crossing the body, wing, or tail.

BENTHIC The lowest level of the ocean, which is inhabited by organisms living in close relationship with the ground. These organisms are called benthos or benthic organisms.

BIVALVES *Molluscs*, such as clams, mussels, and oysters, which have a shell made up of two hinged halves. Most bivalves move slowly or do not move at all, and are *filter-feeders*.

BRACKISH Saltier than fresh water, but less salty than ocean water.

BRACT A leaf-like organ at the base of a flower stalk.

BRACTEOLE A small leaf-like organ at the base of secondary branches of a flower stalk.

BROWNFIELD SITE Previously developed land, now disused and often rich in wildlife.

BRYOZOANS *Filter-feeding* colonial animals that live attached to surfaces, such as seaweed *fronds*, either as flat sheets or as tufty, plant-like growths. Sometimes called "moss animals".

BUR A fruit or seed with a rough or prickly coating.

BYSSUS THREADS The long, strong but fine, silky filaments secreted by some *bivalve molluscs*, by which they attach themselves to a rock or other hard *substrate*.

CALCAREOUS Consisting of or containing calcium carbonate.

CALYX (pl. CALYCES) The collective name for the *sepals* (usually green) which form the outer whorl of a flower.

CARPEL In the female part of a flower, the ovary with a *stigma* attached, which develops after fertilization into a seed.

CATKIN An unbranched and often pendulous flower cluster that has either male or female flowers.

CEPHALOPODS A group of swimming *molluscs* that includes squid, cuttlefish, octopuses, and nautiluses. They have large brains and demonstrate complex behaviour.

CERATA Distinctive, fleshy outgrowths on the back and sides of the body of a sea-slug.

CETACEAN A marine mammal of the group that includes whales, dolphins, and porpoises.

CHEVRON A colour mark or pattern in the shape of a "V".

CHITINOUS Composed of a horny substance (chitin), which forms part of the protective outer coating of many invertebrates, especially insects and *crustaceans*.

CHLOROPHYLL The green pigment of plants and seaweeds that allows them to make their own food by using the energy of the sun.

CHORDATES The *phylum* of animals that includes vertebrates and sea-squirts, characterized by having at some stage of their development a supporting rod, a central nervous system along the back, and gill-slits.

CORALLUM The calcified external "skeleton" of a solitary coral.

CORONA The central, trumpet-like part of a daffodil flower.

COVERT A small feather in a well-defined tract, on the wing or at the base of the tail, covering the base of the larger flight feathers.

CRUSTACEANS The most diverse and abundant group of *arthropods* in the oceans. It includes crabs, lobsters, shrimps, *barnacles*, krill, copepods, *isopods*, and *amphipods*. Their jointed appendages are variously modified as claws, legs, swimming organs, or *filter-feeding* devices, depending on the species.

CRUSTOSE A lichen that closely encrusts, and is very difficult to remove from the surface it grows on.

CYATHIUM (pl. CYATHIA) The unique flower form of a spurge, consisting of a whorl of united *bracts*, a series of nectar glands, and one male and usually five female flowers.

DARK PHASE The darker colour form of a bird species.

DEPOSIT-FEEDING Feeding by extracting food particles from mud or other deposits. See also *filter-feeding*.

DETRITUS Fragments of dead organisms and organic waste material, often mixed with sediment or suspended in ocean currents. A detritivore is an animal that feeds on detritus.

DIATOMS A group of plant-like organisms that belong to the same group as algae, and form a major part of the *plankton*. They are single-celled but often grow as chains or colonies. Diatoms secrete intricate cases of silica around themselves.

DIMORPHIC Having two forms: sexually dimorphic means that the male and female of a species look different; otherwise it indicates two colour forms.

DISC FLORET In the Daisy family, a flower in the central part of the *flowerhead*, whose petals are fused into a tube.

DORSAL Relating to the back or upper surface of an animal.

DRUPE A fleshy fruit whose seeds are surrounded by a tough coat.

DUNE SLACKS Damp depressions, sometimes with permanent water bodies, found within a sand dune system.

ECHINODERMS A major group (*phylum*) of marine invertebrates that includes starfish, brittle- stars, sea-urchins, sea lilies, sea-cucumbers, and sea daisies. Echinoderms have bodies arranged in parts, rather like the spokes of a wheel, with chalky protective plates under their skin. They use a unique system of hydraulic "tube feet" for moving, or for capturing prey, or both.

EPIPHYTE An organism that grows on the leaves, stems, or other structures of a plant, but is not parasitic on it. See also *parasite*.

ESTUARY The mouth of a large river. Used more broadly, the term includes any bay or inlet where sea water becomes diluted with fresh water.

EYEPATCH In a bird, an area of colour around the eye, often in the form of a "mask", broader than an *eye-stripe*.

EYE-RING A more or less circular patch of colour, usually narrow and well-defined, around the eye of a bird.

EYESPOT A marking on the wing of a moth or a butterfly that resembles an exaggerated mammal or bird eye; its function can be to alarm and deter a potential predator.

EYE-STRIPE A stripe of distinctive colour running in front of and behind the eye.

FILTER-FEEDING Feeding by collecting and separating food particles from the environment. When the food particles are suspended in water, it is also called suspension-feeding. See also *deposit-feeding*.

FLORET One of a group of small or individual flowers usually clustered together to form a *flowerhead*.

FLOWERHEAD A cluster of *florets*.

FOLIOSE Describes a lichen that is divided into leafy lobes, lying flat on the *substrate*, but only loosely attached to it.

FORM A term applied to certain species that occur in two or more different colour variations across the species' range.

FROND The leaf-like structure, often divided or branched, of a fern or seaweed.

FRUTICOSE Describes a bushy lichen, with erect or drooping lobes, attached to its *substrate* only at the base.

GASTROPODS The group of *molluscs* that includes snails, slugs, and pteropods (sea-butterflies).

GENUS (pl. GENERA) A category in classification, grouping together closely related species, whose relationship is indicated by the first part of the scientific name, e.g. *Phoca* in *Phoca hispida*.

GLOBOSE Having a spherical, globular shape.

GROYNE An artificial barrier built down a beach and into the sea to hinder transport of materials by currents.

HABIT The shape of a plant.

HERMAPHRODITE An animal that is both male and female. Animals that are both sexes at once are called simultaneous hermaphrodites. Others start as males, then become females, or vice versa. Some species change sex repeatedly.

HOLDFAST A root-like structure that anchors a seaweed to rocks, but does not absorb nutrients like a true root.

HYDROTHECAE The cup- or flask-like structures within which the individuals of many colonial *hydrozoans* reside.

HYDROZOANS Any animal of the class Hydrozoa of the *phylum* Cnidaria. Includes solitary and colonial forms; some colonial forms are attached to a *substrate*, and are branched and seaweed-like, others are floating and jellyfish-like.

IMMATURE In birds, not yet fully adult or able to breed. There may be several identifiable plumages during immaturity, but many small birds are mature by the first spring after they have fledged.

INFLORESCENCE The flowering portion of a plant stem, including flowers, *bracts*, and branches.

INTERTIDAL Defines the area of the shore between the high- and low-water mark.

INVOLUCRE A whorl of bracts below a *flowerhead*, appearing like a *calyx*.

ISOPODS A group of *crustaceans* that usually have flattened bodies. Isopods are mainly marine, but also include the land-living woodlice.

JUVENILE A bird in its first plumage, that in which it makes its first flight, before its first *moult* in the autumn.

KELP One of the group of large, leathery, brown seaweeds, often dominant around the extreme low-water mark of rocky shores.

LAGOON A stretch of coastal water almost cut off from the sea by a spit or other barrier; also, the shallow water within the ring of an atoll.

LARVA The young stage in the life cycle of some insects that follows on from the egg but precedes the *pupa*. In butterflies and moths, it is often popularly referred to as the caterpillar.

LEAFLET One of the divisions that make up a compound leaf.

LIP In flowers, a protruding petal, as in members of the Orchid and Mint families.

LOCAL Having a distribution that is geographically restricted, be that as a continuous range or in isolated pockets.

MANTLE The dorsal body wall covering the main body, or visceral mass in a mollusc.

MERICARP A one-seeded portion of a fruit formed by splitting from the rest.

MICROTIDAL Having a low tidal range, typically with less than 2m difference in height between high and low tide; especially a feature of enclosed or semi-enclosed water bodies.

MIDRIB The primary, usually central, vein of a leaf or *leaflet*.

MOLLUSCS A major group (*phylum*) of invertebrate animals that includes the *gastropods*, *bivalves*, and *cephalopods*. Molluscs are soft-bodied, and typically have hard shells, although some subgroups have lost the shell during their evolution.

MOULT In birds, the shedding and renewing of feathers in a systematic way; most birds have a partial moult and a complete moult each year.

NITROPHILOUS An organism that thrives in areas enriched with nutrients, especially nitrogen. Examples include bird perching sites or rotting driftlines.

OPERCULUM A horny scale that is used to close the opening of many whorled snails; also the flap, which covers the gills of a fish.

OPPOSITE Borne in pairs on opposite sides of a stem.

OSCULUM The mouth-like opening in a sponge, used to expel water.

OVIPOSITOR The part of the abdomen of a female insect that is used to lay eggs.

PALMATE A leaf consisting of three or more leaflets arising from the same point.

PALPS Sensory appendages associated with the mouth of an invertebrate, and concerned with taste and smell.

PAPPUS A tuft of hair on the fruit of a plant, which aids in its dispersal.

PARAPODS The side-extensions on the segments of a bristleworm, often a mixture of fleshy structures and tufts of bristles.

PARASITE An organism that feeds upon another living organism, usually to the latter's detriment.

PAROTID GLAND One of a pair of wart-like glands located behind the eyes in many amphibians, particularly conspicuous in toads. It may produce a noxious secretion.

PEA-FLOWER A flower, usually from the Pea family, with *sepals* fused into a short tube, and with a usually erect upper petal, two wing petals (lateral petals), and two keel petals (lower petals curved like keel of a boat).

PERIANTH The collective term for petals and *sepals*, the structures which surround the reproductive parts of a flower.

PERIOSTRACUM The horny or papery, sometimes coloured, outer covering of some *bivalve molluscs*.

PETIOLE The stalk of a leaf, or the narrowed waist of a wasp.

PHYLUM The highest-level grouping in the classification of the animal kingdom. Each phylum has a unique basic body plan. *Molluscs, arthropods*, and *echinoderms* are examples.

PINNATE Describes a leaf divided into *leaflets* arranged in two rows along a common axis. Pinnately lobed leaves have lobes, rather than leaflets, arranged in this manner.

PLANKTON Marine or freshwater organisms, living in open water, that cannot swim strongly and so drift with the currents. Although small life forms dominate, larger creatures, such as jellyfish, are also planktonic.

PLEUROCARPOUS Describes a moss, usually creeping or prostrate in form, in which the fruit capsules arise on a short side branch.

POLYP One of the two main body forms of cnidarians (an invertebrate bearing tentacles around its mouth). An anemone or coral is a polyp. Polyps are typically tubular and attached to a surface at their base.

PREHENSILE Able to curl around objects and grip them.

PROBOSCIS The coiled but extensible "tongue" of a butterfly or moth.

PRONOTUM In an insect, the *dorsal* cover of the first segment of the *thorax*.

PTEROSTIGMA A coloured panel near the front edge of insect wings.

PUPA The stage in the life cycle of some insects that follows on from the *larva,* and from which the adult insect emerges.

RAY/RAY FLORET The outer, distinctively flattened flower of a daisy-type *flowerhead*.

RECURVED Curved backwards or splayed out.

RHIZOIDS Root-like anchoring structures on various lower plants, including some seaweeds, mosses, and lichens.

RHIZOME A thickened stem (usually underground), which serves as an organ for storing food.

ROSTRUM The tubular, slender sucking mouthparts of some insects. The prolonged part of the head of weevils and scorpionflies.

SEEPAGE ZONE A zone in which fresh water issues from an underlying rock on a broad front, not concentrated in runnels or discrete flows.

SEPAL The usually green parts of a flower surrounding the petals, collectively called the *calyx*.

SILICULA A fruit of the Cabbage family, less than three times as long as broad, and often rounded.

SILIQUA A fruit of the Cabbage family, long and linear or pod-like.

SIPHON In *molluscs*: a fleshy, tubular extension of the body that aids the flow of oxygenated sea water to the gills or sometimes transports food particles for filtering. *Cephalopods* use their siphons for jet propulsion.

SONG-FLIGHT In birds, a special flight, often with a distinctive pattern, combined with a territorial song.

SORUS (pl. **SORI**) A distinct group of minute, *spore*-producing structures on the *frond* of a fern.

SPICULES The structures, made of silica or calcite that stiffen the body of a *sponge*.

SPIKE An unbranched flower cluster, with unstalked flowers.

SPIRE The coiled apex of a snail-like *mollusc*.

SPONGES A large group (*phylum*) of marine animals with a very simple structure, which feed by creating currents through their bodies and filtering small particles from the water. They have no muscles or nerve cells, and sometimes no symmetry.

SPORE A tiny structure produced (usually in large quantities) by non-flowering plants, fungi, and some protists (unicellular organisms), from which a new individual can grow. Spores are much smaller than seeds and usually produced asexually.

SPRING-LINE A zone in which fresh water issues from an underlying rock in distinct, dicrete flows, often under pressure.

SPUR A hollow, cylindrical or pouched structure projecting from a flower, usually containing nectar.

STAMEN Male part of a flower, composed of an *anther*, normally borne on a stalk (filament).

STIGMA The female part of the flower that receives the pollen.

STIPULE A leaf-like organ at the base of a leaf stalk.

STOLON The creeping, horizontal stem of a spreading plant, on or just under the surface of the soil.

STRIATIONS Fine lines forming a texture on the surface of a shell or other structure.

STYLE In flowers, the part of the female reproductive organ that joins the ovary to the *stigma*.

SUBLITTORAL Relating to the coastal marine environment, below the low-water mark.

SUBSPECIES A race; a recognizable group within a species, isolated geographically but able to interbreed with others of the same species.

SUBSTRATE The ground material (rock, shingle, sand, or mud) on or in which a plant or animal lives.

SUPRATIDAL An area or habitat above the reach of the highest tides.

SYMBIOSIS Different species living in an association that brings mutual benefit.

TAP-ROOT The main root of some plants which grows vertically downwards, and has small side rootlets; often swollen.

TERNATE A compound leaf divided into three more-or-less equal leaflets; in twice-ternate, the leaflets themselves are divided into three parts.

TEST The shell of a sea-urchin.

THALLUS The body structure of a lichen, containing both fungal and algal cells.

THORAX The middle section of an adult insect's body, and to which the legs and wings are attached.

TRIFOLIATE A leaf made up of three distinct *leaflets*.

TRINGULIN The small, active early-stage larva of an oil-beetle.

TUBERCLE A raised, wart-like structure on the surface of an *arthropod*.

UMBEL A flat-topped or domed flower cluster with all the flower stalks originating at the same place.

UMBILICUS A navel-like depression in the centre of a flat leaf or similar plate-like plant structure.

UMBO (adj. **UMBONATE**) A bump or hump, near the hinge, on the valves of many *bivalve molluscs*.

UNDERWING The underside of a wing, usually visible only in flight or when a bird is preening.

UROPOD The leg-like appendages arising from the tail end segments of an *amphipod crustacean*.

VARIEGATED Having more than one colour; usually used to describe leaves.

VERRUCARIA ZONE The black lichen zone just above the high-water mark on a rocky coast.

VOLVA A sac-like bag or swelling at the base of the fruiting bodies of some fungi.

WINGBAR A line of colour produced by a tract of feathers or feather tips, crossing the closed wing of a bird and running along the spread wing.

WINGPIT In a bird, a group of feathers – the axillaries – located at the base of the *underwing*.

XANTHORIA ZONE The yellow/orange lichen zone on a rocky coast, lying immediately above the black *Verrucaria zone*.

ZOOID An individual in a colony of interconnected animals, such as *bryozoans*. The term is not applied to colonial coral animals, which are termed *polyps*.

Index

D

E

F

G

H

I

J

K

L

Q

R

S

T

U

V

W

X

Y

Z

Acknowledgments

The author would like to thank his wife, Maureen, for her constant help and support. Dorling Kindersley would like to thank Miezan Van Zyl and Tamlyn Calitz for editorial assistance, and the following for their kind permission to reproduce their photographs:

(Key: a-above; b-below/bottom; c-centre; f-far; l-left; r-right; t-top)

Alamy Images: Enrique R Aguirre Aves 2-3; allOver photography 147bl; Arco Images 9crb, 43clb, 111cra; Blickwinkel 79cra, 152br, 165b; K Byrne 157bl; Nigel Cattlin 56br; David Chapman 136ca, 214ca; Bill Coster 186bl; CuboImages Srl 165c; FLPA 214cb; Bob Gibbons 48cra; Steven J Kazlowski 206cla; Nature Picture Library 83bl, 83br, 84cla, 158cla; David Osborn 33bc, 33bl, 33br; Wolfgang Polzer 118bl; Anestis Rekkas 152t; Tim Ridley 122br; Wildchromes 172cla; Tim Woodcock 83cra; **Algaebase.org:** Colin Bates 85clb; Dr Felicini 88bc, 88bl; **Roy Anderson:** 53cra; **Ardea:** John Mason 23clb, 140ca; **Photo Biopix.dk:** J C Schou 15bc, 46br, 55cl, 80bl, 81c, 84cra, 85bl, 130tr, 132tr, 133bl, 138bl; N Sloth 143cb, 160c; **Dan Bolt:** 94cra, 109cra, 119bl, 123cla, 160bl; **California Fungi/ Michael Wood:** 74cra; **University of California:** CalPhotos/Luigi Rignanese 127br; **Corbis:** Niall Benvie 79bl; Eric Crichton 76br; Chinch Gryniewicz / Ecoscene 30crb, 73bl; Warren Jacobi 30bl; **Graham Day:** 15br, 43br; **DK Images:** Chris Gibson 67c; **Ecomare (www.ecomare.nl):** Sytske Dijksen 129br; **David Fenwick (www.aphotoflora.com):** 11crb, 19cra, 23bl, 25bl, 25clb, 26bc, 26c, 30cra, 37cla, 37cra, 41br, 41clb, 47cl, 50clb, 50ftl, 53br, 57cla, 59cl, 59cra, 61bl, 61br, 65bl, 65crb, 65fbr, 66tl, 73cla, 73tl, 79cla, 80cra, 81br, 83cla, 84cb, 95cla, 101cla, 101cra, 104br, 105bl, 108cr, 108cra, 110bl, 110br, 112bl, 112br, 113cra, 114cla, 114clb, 115br, 117bl, 117cb, 118cr, 118tr, 119cra, 120cla, 121cra, 122cl, 122cra, 124br, 126bl, 126cra, 127cra, 131c, 131cra, 132b, 134br, 134crb, 135tl, 158br, 159br, 159clb, 162ca, 211t; **Flickr / Derek Haslam:** Derek Haslam 100cla; **Flickr / Lynden Schofield:** 26bl, 94cla; **Flickr / Taco Meeuwsen:** Taco Meeuwsen 104cla; **Flickr / Tim Riches:** 4-5; **Flickr / Vidar Aas:** 91br, 96br; Vidar Ass 54bl; **Flickr.com:** 6l, 12t, 123bl, 150bl, 209b, 209clb; Jim Anderson 90bc, 103cra, 117br, 122cla; Ryan Backman 121ca; Jennifer Batten 70c; Roger Butterfield 52br; CA Floristics 68br; Brian Chan 142cla, 142cra; Mark Craig 107bl; Cyprid 91cr; Mark Ellie 138cla; Elisabeth Guegan 110cla, 110cra, 117cla, 161bl; Frank Hager 208bl, 212b; Asbjorn Hansen 115tl; Jim Howard-Birt 212cr; Henrik Kibak 113bl; Keith Marshall 185clb; Heather McCallum 88c; Ellen Mikkelsen 211cr; Thomas Palmer 28ca; Ferran Pestana 172cra, 184br; Andy Phillips taken at Hasting Country Park Nature Reserve 147cra; Joey Ramone 76clb; Rebecca Robinson 110cl; Stephanie Rousseau 208br, 214b; Ron Wolf 112cla; **Floral Images / John Crellin:** 10cra, 22bl, 22br, 22clb, 37tl, 51bl, 51cl, 51cla, 51crb, 59bl, 59br, 73br, 77cla, 300; **FLPA:** Robin Chittenden 49cl, 61cla; Foto Natura 131clb; Bob Gibbons 40ca; Panda Photo 206br; Peter Reynolds 83cl; D P Wilson 86crb, 102c; **Dr Daniel L Geiger:** 141bl; **Getty Images:** Sisse Brimberg 168t; **Chris Gibson:** 8b, 8t, 9tl, 11cl, 11tl, 14cl, 14tl, 15bl, 15cr, 15crb, 15tr, 16bc, 16cl, 16tl, 17cla, 17cr, 17cra, 17tl, 17tr, 18tl, 19cl, 19cr, 19crb, 19tr, 20ca, 20cl, 20tl, 21br, 21cr, 21tr, 22cl, 22tl, 23br, 23cr, 23tr, 24tl, 25br, 25cr, 25tr, 26tl, 27br, 27cr, 27t, 28cl, 28tl, 29cr, 29cra, 29tr, 30cl, 30cla, 30tl, 31tr, 32cl, 32tl, 33crb, 33tr, 34bc, 34bl, 34tl, 35cr, 35tr, 36tl, 37cr, 37tr, 38cr, 38tl, 39cr, 39tr, 40cl, 40cla, 40tl, 41cr, 41tr, 42tl, 43cb, 43cr, 43tr, 44b, 44cra, 44tl, 44tr, 45tr, 46cl, 46cla, 46tl, 47bl, 47tr, 48bl, 48br, 48c, 48cl, 48crb, 48tl, 49bc, 49bl, 49ca, 49cr, 49tr, 50bl, 50ca, 50cl, 50cla, 50cra, 50tl, 51br, 51cr, 51cra, 51tr, 52bl, 52cl, 52cra, 52tc, 52tl, 53bl, 53cl, 53cr, 53tr, 54cl, 54cla, 54crb, 54tl, 55bl, 55cla, 55cr, 55cra, 55tr, 56cl, 56clb, 56cra, 56tl, 56tr, 57cr, 57cra, 57tr, 58cl, 58tl, 59cr, 59tr, 60bl, 60br, 60cl, 60cla, 60tl, 61cr, 61crb, 61tr, 62tl, 63bl, 63br, 63cla, 63cr, 63cra, 63tr, 64b, 64bc, 64c, 64tl, 65ca, 65cl, 65cra, 65tr, 66bc, 66cl, 66cla, 66cra, 66fcl, 66ftl, 67tr, 68bl, 68cl, 68cra, 68tl, 69cr, 69tr, 70bc, 70bl, 71bl, 71cla, 71crb, 71tr, 72tl, 73cr, 73tr, 74cl, 74tl, 75b, 75tr, 76cl, 76cla, 76tl, 77br, 77cr, 77tr, 78br, 78tl, 79cr, 79tr, 80br, 80cl, 80tl, 81cla, 81cr, 81tr, 82tl, 83tr, 84cl, 84tl, 85br, 85cr, 85tr, 86cl, 86tl, 87cr, 87tr, 88cl, 88cra, 89ca, 89cr, 89cra, 89tr, 90br, 90fbr, 91crb, 91tr, 92cl, 92tl, 93cr, 93tr, 94cl, 94tl, 95cr, 95tr, 96cl, 96clb, 96tl, 97tr, 98b, 98cr, 98tl, 99bl, 99cla, 99cr, 99cra, 99tr, 100cl, 100tl, 101cr, 101tr, 102tl, 103bl, 103cr, 103tr, 104cl, 104tl, 105c, 105tr, 106cl, 106tl, 107cr, 107tr, 108b, 108tl, 109cr, 109tr, 110clb, 110tl, 111br, 111cr, 111tr, 112cl, 112tl, 113cr, 113tr, 114br, 114cl, 114cra, 114tl, 115cr, 115tr, 116b, 116tl, 117cr, 117tr, 118cl, 118cla, 118tl, 119cr, 119tr, 120ca, 120cl, 120cra, 120tl, 121cr, 121tr, 12 2clb, 122tl, 123cr, 123tr, 124bl, 124clb, 124tl, 125cr, 125tr, 126cl, 126cla, 126tl, 127cr, 127tr, 128cl, 128tl, 129crb, 129tr, 130clb, 130tl, 131cla, 131crb, 131tr, 132cl, 132cla, 132clb, 132tl, 133tr, 134bl, 134cl, 134tl, 135cr, 135tr, 136br, 136clb, 136tl, 137cr, 137tr, 138cl, 138tl, 139cr, 139tr, 140cl, 140tl, 141cr, 141tr, 142cl, 142tl, 143cr, 143cra, 143tr, 144br, 144cl, 144tl, 145br, 145cr, 145tr, 146bc, 146bl, 146cl, 146tl, 147clb, 147cr, 147tr, 148cl, 148cla, 148tl, 149cr, 149cra, 149tr, 150cl, 150tl, 151cr, 151tr, 153tr, 154cl, 154tl, 155tr, 156cl, 156tl, 158cl, 158tl, 159cr, 159tr, 160cl, 160tl, 161cr, 161tr, 162cl, 162tl, 163cr, 163tr, 164cl, 164tl, 165tr, 166cl, 166tl, 167tr, 169tr, 170cl, 170tl, 171cr, 171tr, 172cl, 172tl, 173tr, 174br, 174tl, 175cr, 175tr, 176cl, 176tl, 177tr, 178tl, 179tr, 180cl, 180tl, 181cr, 181tr, 182tl, 183cr, 183tr, 184cl, 184tl, 185tr, 186tl, 187cr, 187tr, 188tl, 189cr, 189tr, 190cl, 190tl, 191cr, 191tr, 192tl, 193cr, 193tr, 194cl, 194tl, 195tr, 196tl, 197cr, 197tr, 198cl, 198tl, 200cl, 200tl, 201cr, 201tr, 202cl, 202tl, 203tr, 204tl, 205cr, 205tr, 206cl, 206tl, 207tr, 208fbl, 209ca, 209tr, 210tl, 212cla, 212crb, 212tl, 214cr, 214tl; **Maureen Gibson:** 103cla; **Alan Gilbertson:** 203tl; **Guido & Philippe Poppe - Conchology, Inc. 2007:** G. & Ph. Poppe 125cla; **Michael Guiry:** 70fbr, 78bl, 82c, 85cla, 85cra, 86cla, 86cra, 87bl, 87br, 87cla, 87cra, 89b, 93br, 119cb; **Peter Harvey:** 151bl, 151cr, 151cra; **Keith Hiscock:** 86br; **Image Quest 3-D:** Valdimar Butterworth 96cla; **iStockphoto.com:** David Acosta Allely 40cra; Angelafoto 111cla; Pathathai Chungyam 127cl; Ruud de Man 28cb; Peter Elvidge 208t; Shirley Friedman 211br; Renate Micallef 212cl; Stephen Morris 47c; Giacomo Nodari 133c; Simon Phipps 210b; Giovanni Rinaldi 9bl; Jamie Roset 75ca; Ben Slater 157b; Dave White 79br; **Lars Iversen:** 150cl; **Roger Key:** 142bl, 142br; **Peter Krogh:** 148tr; **Arne Kuilman:** 91cl, 91cra, 112cra; **Rosanna Milligan:** 124cr; **Jeff Moore (jeff@jmal.co.uk):** 113br; **Mossimages /Des Callaghan:** 71br, 71cr; **Natural Visions:** 71cra, 82b, 101bl, 101br, 104bl, 118br, 119cla, 121br, 129cla, 134cra, 140br; **naturepl.com:** Jane Burton 139cla; **Karen Nichols:** 149bl, 149cb; **NOAA:** 123crb; **Julia Nunn:** 106ca, 120bl, 128bl, 210br, 210cla; **Filip Nuyttens:** 94br, 95bl, 95br, 104cra, 107cla, 136c, 211tr, 213b, 213tl; **OSF:** Bob Gibbons 72cr; **Papiliophotos:** Peter Tatton 109bl; **PBase:** Nicholas Bond 181cb; Stefano Guiducci 10crb, 189ca; Daniele Occhiato 198cr; Hans van Rafelghem 76cra; **PBase / Tom Murray:** 143br; **Photoshot / NHPA:** David Higgs 8crb, 158tr; Roy Waller 210cr; **Bernard Picton:** 90bl, 90t, 92bl, 92br, 92cla, 92cra, 93cl, 93cra, 94bl, 95cra, 96cra, 97c, 99br, 100br, 100cra, 106br, 107br, 109br, 109cla, 111bl, 112clb, 122bl, 124cl, 124cla, 126br, 127cb, 127clb, 128br, 128cr, 128cra, 129cra, 130c, 130cla, 131bl, 131cb, 135b, 135cra, 137ca, 138br, 161cb; **Premaphotos Wildlife:** 100bl, 103br, 113cla; **Copyright Trustees of the Royal Botanic Gardens, Kew:** 46bl; **rspb-images.com:** Nigel Blake 189crb; Mark Hamblin 201ca; Chris Knights 206cra; Mike Lane 1, 179c; **Thomas Schoepke (www.plant-picture.com):** 29cla, 67bc, 67bl; **Science Photo Library:** Bob Gibbons 56cla; **Sierradelta / Stephen Dent:** 117cra; **Johannes F Skaftason:** 143ca; **Robert Thompson:** 148bl, 148br, 149cla, 149tl; **ukwildflowers.com / Peter Llewellyn:** 70br, 73cra; **US Geological Survey:** Woods Hole 106bl; **www.UWPhoto.no:** Rudolf Svensen 152bc, 158bl, 161ca, 162cb; Erling Svenson 84b, 121bl, 160ca, 160cb; **Cintaras Vaidakavicius:** 206bl; **Nicolas J Vereecken:** 144bl; **Wikimedia Commons:** 35bl, 213c; Hans Hillewaert 9br, 129bl, 130bl; Juergen Howaldt 66br; Carsten Niehaus 16bl, 20cla; Arnold Paul 139b; Roberto Petruzzo-Latina 74cla; Adrian Pingstone 168bc, 184bl; Julio Reis 9cra, 72bc, 102b; Stemonitis 81bl; Kurt Stuerber Collection / Prof.Dr. Otto Wilhelm Thome, Flora von Deutschland, Osterreich und der Schweix 1885, Gera, Germany 55br; Malene Thyssen 213cra; Tigerente 38cla; Andreas Trepte 123cra; Velela 65br; M Violante 125clb; Yummifruitbat 30br; **Mark Wright:** 74bl; **www.uklichens.co.uk:** 77cra